THE GROVER E. MURRAY STUDIES IN THE AMERICAN SOUTHWEST

Also in the series

Brujerías: Stories of Witchcraft and the Supernatural in the American Southwest and Beyond, by Nasario García

Cacti of Texas: A Field Guide, by A. Michael Powell, James F. Weedin, and Shirley A. Powell

Cacti of the Trans-Pecos and Adjacent Areas, by A. Michael Powell and James F. Weedin

Cowboy Park: Steer-Roping Contests on the Border, by John O. Baxter

Dance All Night: Those Other Southwestern Swing Bands, Past and Present, by Jean A. Boyd

Deep Time and the Texas High Plains: History and Geology, by Paul H. Carlson

From Texas to San Diego in 1851: The Overland Journal of Dr. S. W. Woodhouse, Surgeon-Naturalist of the Sitgreaves Expedition, edited by Andrew Wallace and Richard H. Hevly

Grasses of South Texas: A Guide to Identification and Value, by James H. Everitt, D. Lynn Drawe, Christopher R. Little, and Robert I. Lonard

In the Shadow of the Carmens: Afield with a Naturalist in the Northern Mexican Mountains, by Bonnie Reynolds McKinney

Javelinas: Collared Peccaries of the Southwest, by Jane Manaster

Little Big Bend: Common, Uncommon, and Rare Plants of Big Bend National Park, by Roy Morey

Lone Star Wildflowers: A Guide to Texas Flowering Plants, by LaShara J. Nieland and Willa F. Finley

Myth, Memory, and Massacre: The Pease River Capture of Cynthia Ann Parker, by Paul H. Carlson and Tom Crum

Pecans: The Story in a Nutshell, by Jane Manaster

Picturing a Different West: Vision, Illustration, and the Tradition of Austin and Cather, by Janis P. Stout

Seat of Empire: The Embattled Birth of Austin, Texas, by Jeffrey Stuart Kerr

Texas, New Mexico, and the Compromise of 1850: Boundary Dispute and Sectional Crisis, by Mark J. Stegmaier

Texas Quilts and Quilters: A Lone Star Legacy, by Marcia Kaylakie with Janice Whittington

The Wineslinger Chronicles: Texas on the Vine, by Russell D. Kane

WILLA F. FINLEY LASHARA J. NIELAND

LAND OF ENCHANTMENT
Wildflowers

A GUIDE TO THE PLANTS OF NEW MEXICO

TEXAS TECH UNIVERSITY PRESS

This book is typeset in Stempel Schneidler Std. The paper used in this book meets the minimum requirements of ANSI/NISO Z39.48-1992 (R1997). ♾

Design, composition, and project management by Books By Design, Inc.
Cover design by Barbara Werden

Library of Congress Control Number: 2013940185
ISBN (paper): 978-0-89672-822-6

Printed in Korea

13 14 15 16 17 18 19 20 21 / 9 8 7 6 5 4 3 2 1

Texas Tech University Press | Box 41037 | Lubbock, Texas 79409-1037 USA
800.832.4042 | ttup@ttu.edu | www.ttupress.org

To Arnold Clifford,
a consummate naturalist and ethnobotanist
who inspires others with his dedication to preserving the
knowledge of New Mexico's plants

ACKNOWLEDGMENTS

We are grateful to many people who made the journey toward completion of this book a pleasant and exciting adventure.

The professionals at Texas Tech University Press expertly directed us at each step along the way. Robert Mandel and Judith Keeling encouraged us to write a follow-up to our first book, *Lone Star Wildflowers: A Guide to Texas Flowering Plants*. Robert expedited the process, allowing us to make this work a more speedy reality than the first book. Judith, in her gentle and kindly way, gave us ideas on how to give this current work a New Mexico flavor with an emphasis on Native American interactions with wild plants. Judith, along with Joanna Conrad, offered a wealth of technical guidance. Kasey McBeath and Kellyanne Ure graciously and patiently helped us as we used their office space at Texas Tech during the photo-editing process. They impressed us with their work ethic and their creative abilities in book design and production. We appreciate the efforts of Jada Rankin and John Brock in getting this book into readers' hands. When we called TTUP on numerous occasions with a variety of questions, Isabel Williams and LaTisha Roberts were sources of information given in a helpful manner.

We also want to express our gratitude to Margaret Merritt, who granted us access to her lovely ranch in southeast New Mexico, where we also enjoyed the hospitality of Dan and Bev Bell. In the opposite direction, we were welcomed by Heidi and Abdel Berrada in the Four Corners area, where we enjoyed photographing plants in the hills and canyons near their home. Joe and Beverly Campbell, and Joe's mother and Willa's aunt, Sissy Campbell, also welcomed us when we visited the Las Cruces area.

We want to thank Charlie Shanks, LaShara's father, for giving us access to his collection of Southwest artifacts of plant materials that played a key role in Native American life for centuries. While visiting White Sands National Monument, we were especially touched by the enthusiasm the park rangers showed for our project and the help they provided in identifying several species. The staff at Salmon Ruins were similarly encouraging, and we gathered some excellent research material in their well-stocked bookstore. The Native American vendors at the weekly flea market in Gallup, New Mexico, were friendly and informative, and we particularly want to thank Deanna Tso, who explained how certain plants are used in her Navajo culture.

We especially want to thank Robert Johnson of the Navajo Nation Museum in Window Rock, Arizona, who visited with us at length regarding the use of plants by Native Americans in the Southwest.

Our deepest gratitude also goes to Arnold Clifford, a Navajo ethnobotanist, who took us into the spectacular backcountry of the Navajo Nation in the rugged Four Corners area, sharing his deep knowledge of the plants and explaining their traditional uses. We appreciate his insights and expertise so very much.

CONTENTS

LAND OF ENCHANTMENT WILDFLOWERS

INTRODUCTION

It is not without reason that New Mexico is called the Land of Enchantment. From the stark delights of the deserts to the soaring mountain ranges and the peaceful, windswept grasslands, the varied vistas capture the mind and heart, drawing them into a realm of beauty that is difficult to match. Intertwined in the experience is the realization that the region's plants are an integral part of the experience that is uniquely New Mexico.

Our nation's fifth-largest state, New Mexico, sits at the intersection of the Rocky Mountains, the Great Plains, and the Sonoran and Chihuahuan Deserts. The some 4,100 species of plants that inhabit and support these varied ecosystems are major players in the story of the state, and it is our purpose to give a more complete account of the flora than a typical field guide might offer.

The parks system, national forests, and historic sites in New Mexico are second to none and provided us with ample opportunities for photography and research. In trekking through the Organ Mountains, Bandelier National Monument, Salmon Ruins, Aztec Ruins National Monument, White Sands National Monument, Carlsbad's Living Desert State Park, City of Rocks State Park, Rio Grande Gorge State Park, Navajo Lake State Park, and many others, we gained a new appreciation for the state's beautiful and varied plant communities. Some of our richest experiences were within the tribal lands of the Navajo Nation near the Four Corners area, as well as in the lands of other Native Americans scattered throughout the state. New Mexico has many informative museums, such as the Indian Pueblo Cultural Center in Albuquerque and the Museum of Indian Arts and Culture in Santa Fe, where visitors can explore New Mexico's rich cultural heritage, including the plants that were, and are, a part of Native Americans' lives.

Lone Star Wildflowers: A Guide to Texas Flowering Plants, our previous book, influenced the design and content of this field guide. Helpful feedback from its readers from all walks of life and with all levels of experience with wildflowers informed our choices in content and format as we developed this field guide.

This book includes the following features:

- *Accessible format,* with the plants grouped by the *color* of their flowers. Some flowers intergrade, such as the pinks and purples, so if you don't find your specimen in one category, check the other. We have provided *cross-references* to link plants with similar properties, particularly those within the same botanical family.

- *Name origins* of both the common and scientific names, with special notes on the botanical explorers and plant collectors the names commemorate, especially during the golden era of the 1800s.
- *Descriptions and photographs* of not only *flowers* but also buds, leaves, seeds, spines, and particularly *fruits*, which are as varied and creatively inspired as the blooms. They range from tiny, dry types to the large, fleshy ones that most people associate with the word *fruit*. These seed-bearing, ripened ovaries are adorned with all manner of fuzz, wings, and hooks.
- *Complete coverage* of all the state's vegetation zones, in all four quadrants and at all altitudes. We traveled throughout the state during the spring, summer, and fall to photograph plants in *various life stages*.
- *Historical and modern uses* of the plants for *food, medicine, ceremonies,* and other applications, along with *archaeological findings*. We include photographs of *artifacts* relating to the plants' historical uses.
- *Toxins* and other special chemical compounds, some of which are commercially valuable.
- *Habitat information,* including each plant's range within New Mexico, associations in plant communities, flowering and seeding seasons, elevation, soil types, and growth habit.
- *Interactions of livestock, wildlife, and insects with plants.* Insects photographed with flowers are identified for ease of reference.
- *Landscaping* uses for New Mexico's USDA hardiness zones 4–8, with notes on soil, water, and light requirements, as well as *deer resistance* and ability to attract *butterflies* and *moths*. New Mexico is ranked third in the nation for butterfly populations, with approximately three hundred species.
- *Embedded and stand-alone glossaries* explaining technical terms both in context within the reading and summarized in a dedicated glossary for handy reference.

Vegetation Zones

In describing the species, we identify the part or parts of the state in which the plant grows. We often mention the vegetation zone making up the plant's habitat. Here is a brief overview of the zones, which are generally named for the dominant plant species:

- *Riparian* — along rivers
- *Desert* and *Grasslands* (*includes creosote bush/chaparral*) — up to about 6,000 feet elevation

- *Pinyon-Juniper (includes sagebrush)* — about 5,000 to 7,000 feet elevation
- *Ponderosa Pine* — about 7,000 to 9,000 feet elevation
- *Douglas Fir* — about 9,000 to 10,000 feet elevation
- *Spruce-Fir* — about 10,000 to 12,000 feet elevation
- *Alpine* — above about 12,000 feet elevation

Native American Culture

One of the aspects that makes New Mexico very special is its vibrant Native American culture, which reaches back thousands of years. Archaeological findings indicate the foundational role plants played among the state's original inhabitants, and modern Native Americans still utilize much of the flora in traditional ways. In writing the account of each plant, we sometimes found it difficult to determine whether to use past or present tense; for example, many medicinal plants are known today but not actually used anymore. However, a segment of the Native American population still relies on many of the old cures. When we went to the weekly flea market in Gallup, we were fascinated to see a variety of herbal remedies for sale there, demonstrating that the treasure trove of plant lore is still alive and well. In visiting with the vendors, we learned that some people are very scrupulous in collecting medicinal plants, using traditional religious rituals, offerings, and prayers. It is difficult to emphasize strongly enough how much herbal cures rely on spiritual overtones and community support, in addition to the pharmaceutical properties of plants, to effect healing. Indeed, New Mexico's native peoples are in good company when they utilize the ancient knowledge of traditional herbal healers. The World Health Organization has found that more than 75 percent of the world's rural populations rely on herbalists rather than modern medical doctors to meet their health needs.

We have endeavored to convey the historical and modern practical uses of plants by different Native American groups for food, fiber, medicine, construction, crafts, dyes, and ceremonies. We drew upon historical records as well as conversations with the kind and very knowledgeable people we met. Most of our in-depth conversations were with Navajo people, but we made an effort to give equal coverage to other tribes as well, including the following major cultural groups found in New Mexico:

- *Apache*: The *Chiricahua Apache* live west of the Rio Grande, extending into Arizona and Mexico. On the eastern side of the river are the *Mescalero Apache*, who reach into Texas and Chihuahua, Mexico.
- *Isleta*: Their pueblo is twelve miles south of Albuquerque on the west bank of the Rio Grande.
- *Jemez*: Their pueblo is on the Jemez River forty-five miles northwest of Albuquerque.
- *Keresan*: People of this language group live in seven pueblos, with the *Acoma* and *Laguna* located west of Albuquerque and the *Cochiti, San Felipe, Santa Ana, Santo Domingo,* and *Sia* (*Zia*) living to the east.
- *Navajo* (*Diné, Dineh*): These people are the most populous of all the Native American groups in the United States. The *Ramah Navajo* live in western New Mexico, and the *Kayenta Navajo* extend into northeast Arizona.
- *Tewa*: This linguistic group of pueblo tribes has homelands in the vicinity of the Rio Grande near Santa Fe.
- *Zuñi*: Their pueblo is about forty miles southwest of Gallup.

Caution

A strong word of caution is necessary. Many of the plants mentioned in this book are toxic, even to the point of being fatal. Using any of these plants for medicine or food is potentially dangerous, because some edible plants have toxic look-alikes and many medicinal plants can be hazardous. The active chemicals in the plants vary with the season and growing conditions, and it is perilous to undertake a herbal healing without extensive training.

Red Flowers

ASPARAGACEAE (Asparagus Family) (formerly in Agavaceae)

Agave lechuguilla

LECHUGUILLA, SHINDAGGER

When Linnaeus chose the name for *Agave*, he based it on the Greek word *agauos*, which means "illustriously noble" or "admirable." Like many of the names Linnaeus gave, it has many meanings, also referencing the Greek goddess Agave, the daughter of Harmonia. The name reflects the auspicious appearance of the flower stalk, which ascends nine to twelve feet, and is replete with bright flowers during the bloom period from May to June.

Lechuga, Spanish for "lettuce," with the diminutive form *illa*, meaning "little," gives this tough, spiky plant its species and common names, although in truth it's not very much like lettuce. The name "shindagger" seems more appropriate. Indeed, the dense colonies of lechuguilla in the Chihuahuan Desert were one of the perils that Spanish conquistadors complained of, with their animals' feet and legs, as well as their own, suffering greatly from the one- to two-inch spines on the tips of the knee-high leaves.

The flowers emerge into glory when the plant matures at ten to twenty years of age. The flowering stalk begins its life within the "heart" of a rosette of thickly succulent leaves, forming in response to time, light, and nutrients, which, at the right moment, collectively give the signal, "*Agave*, start flowering!" After a few months of splendor, the plant dies, and a number of its clones (see photo) develop to carry the past into the future.

Agave lechuguilla, lechuguilla; field view, in flower.

Agave lechuguilla, lechuguilla; seedpods.

Agave lechuguilla, lechuguilla; clonal group, with dead and young plants.

The color of the tepals (petals and sepals that are identical in appearance) range from yellow to red, with the red filaments of the well-exserted anthers giving the flowers a preponderantly red color. The fruit is oblong and about an inch long.

Lechuguilla grows in rocky limestone soil in several plant communities from 2,800 to 6,200 feet elevation, including grasslands, creosote bush, and pinyon-juniper. An indicator plant for the Chihuahuan Desert, its presence defines the boundaries of this desert, whose northern limits are in southern New Mexico. Reproduction is almost entirely vegetative, with large areas covered by clonal groups. The root system is very shallow, from two to four inches deep, and up to three times wider than the main mass of the clonal group. This allows the plants to absorb significant quantities of water from even brief rainfall events. Another of its drought-tolerance adaptations is the chemical composition of the leaves and roots, which contain large amounts of salts, mucilage, and saponin compounds, all of which prevent dehydration.

Every part of lechuguilla played a useful role in the life of American Indians living in the Chihuahuan Desert. The short inner core, or stem, to which the leaves are attached, contains large quantities of carbohydrates, along with the toxic saponins. Native Americans cooked this "heart" in a pit for two full days to destroy the saponins, then formed the flesh into sweet cakes or fermented it into an alcoholic drink. The cakes were eaten fresh, or dried and stored for months like those of sotol, *Dasylirion wheeleri* (p. 270).

The leaves contain hard fibers, called istle or ixtle, that Indians used to make baskets, rope, sandals, brushes, and clothing. Archaeological excavations in the Lower Pecos Canyonlands have uncovered artifacts indicating the use of lechuguilla fibers at least eight thousand years ago. Today istle is known commercially as Tampico fiber. Extracted by small operators in Mexico, it is used for making rope, brushes, mats, and carpets.

Native Americans used the dried flower stalks for construction of walls and roofs of houses, fishing poles, spear shafts, and firewood.

The saponins in lechuguilla make a mild detergent that gently cleans the hair and skin, as well as delicate fabrics. In the past, the saponin-containing stems and leaves were powdered and used to stun fish for easy capture, while juice from the leaves and stems provided a poison for arrow tips.

One of lechuguilla's saponins is smilagenin, which is receiving some attention today for its use in treating cognitive disorders, including Alzheimer's disease. Another is hecogenin, a raw material in the synthesis of cortisone.

Lechuguilla is lethal to domestic sheep, goats, and cattle; poisoning usually occurs when drought conditions prevail and range condition is poor. Sheep and goats are particularly affected, becoming lethargic, anorexic, and jaundiced; experiencing swollen mucous membranes; and excreting yellow fluid from the eyes and nose.

Mule deer eat the young flower stalks and baby plants with impunity, and collared peccaries (javelinas) consume the tender leaves of the growing point, the base of leaves that are on the outside, and the roots. Pocket gophers also eat the inner core at the growing point. Many small mammals call lechuguilla home, and it provides cover and nesting sites for many species of birds, lizards, and snakes.

Lechuguilla's stately inflorescences make it a distinctive feature in desert xeriscapes. The nectar of lechuguilla flowers is popular with an assortment of bees, beetles, butterflies, wasps, and hummingbirds. Lechuguilla is a larval host and nectar source for the Mary's giant-skipper butterfly (*Agathymus mariae*).

Agave lechuguilla hybridizes with other *Agave* species, including *A. parryi* subsp. *neomexicana* and *A. havardiana*, to create intergrading populations.

ASTERACEAE (Sunflower Family)

Gaillardia pulchella

INDIAN BLANKET, FIREWHEEL

The annual *Gaillardia pulchella* is found throughout New Mexico, while the presence of the perennial *G. aristata* is not as extensive. These two species share many attributes in their habitats and uses.

The genus *Gaillardia* honors the French legislator Gaillard de Charentonneau, who was a benefactor of botanical research and exploration in the 1700s. *Pulchella* is Latin for "beautiful," surely an apt description of this delightful wildflower. *Aristata* refers to the bristles found on the pappus hairs attached to the achenes.

The name "Indian blanket" comes from an old legend about a blanket weaver who wished to express his gratitude to the Great Spirit. The old man used the colors of the sunset to weave the most wonderful blanket he had ever made. When he was ready to pass on to the Great Hunting

Gaillardia aristata, great blanketflower; perennial.

Ground, he asked to be buried in this yellow, red, and brown blanket so that he could present it as a gift to his Maker. To honor the generous weaver, the Great Spirit lavishly blanketed the land with this flower, which displays all the colors of the sunset.

Indian blanket grows in sandy or calcareous soil, as well as in well-drained loamy soil, in several plant communities up to 9,000 feet elevation.

Indian blanket contains antimicrobial compounds and the sesquiterpene lactone gaillardin, which has tumor-inhibiting properties. Several tribes used the roots of Indian blanket to prepare a tea to treat stomach conditions and made a powder of the roots to apply to skin disorders. They also made a diuretic tea for urinary problems. Nursing mothers bathed their sore nipples in Indian blanket tea, and Acoma and Laguna mothers rubbed the plant on their breasts to wean their infants.

Pale green and bright yellow dyes can be extracted from the flower heads, with the final hue depending on the compounds used to fix the colors.

Indian blanket has one of the most memorable faces among wildflowers and is now a well-known drought-tolerant ornamental not only in Southwest landscapes but also in more rain-rich parts of the country. Seed is available commercially in wildflower mixes for different parts of the country. The plant reseeds prolifically.

A number of insects use Indian blanket as a nectar and pollen source, including the butterfly species Edwards' fritillary (*Speyeria edwardsii*) and Uncas skipper (*Hesperia uncas*). Finches eat Indian blanket seeds.

Also shown here is *G. aristata* (great blanketflower), a robust perennial with larger flowers than all its relatives.

For information about another *Gaillardia* species, see p. 53.

CACTACEAE (Cactus Family)

Echinocereus coccineus

CLARET-CUP CACTUS, STRAWBERRY HEDGEHOG CACTUS, KINGCUP CACTUS

Claret-cup cactus, with its strikingly handsome scarlet flowers, is a signature plant of Southwest deserts. The genus name comes from the Greek *echinos*, meaning "spine," combined with *cereus*, or "waxy," which refers to the waxiness of the flowers. The species name, *coccineus*, means "scarlet," recognizing the brilliant orange-red flowers.

Echinocereus coccineus, claret-cup cactus; whole plant.

Claret-cup grows throughout New Mexico in limestone or sandy soil from 4,000 to 9,000 feet elevation, with "family clusters" reaching one to two feet tall and up to four feet wide. The flowers appear from March to June, and the red fruits mature about three months later. The cactus exhibits morphology gradients from the northern to southern parts of its range, particularly in spine number (fewer in the south) and plant size (larger in the south).

All members of the genus *Echinocereus* have juicy fruits that are somewhat smaller than those of the prickly pear (see p. 96), but these fruits are sweeter and have spines that are more easily removed. Native Americans ate the fruits in a variety of ways, including as preserves. Cactus pulp is gel-like and was used as a source of water by the Isleta in emergencies. The pulp was also cooked with sugar to make cakes and candy. The Isleta roasted the stems and placed them on sores to reduce swelling.

Claret-cup is pollinated by hummingbirds. Little green halictid sweat bees also frequent claret-cup, as do honeybees.

For information about another *Echinocereus* species, see p. 232.

FOUQUIERIACEAE (Ocotillo Family)

Fouquieria splendens

OCOTILLO, COACH WHIP

Fouquieria splendens was named for Pierre Fouquier, the nineteenth-century French physician for Charles X and Louis-Philippe. The species name, meaning "shining" or "gleaming," refers to the striking lipstick-red flowers. *Ocotillo* comes to us as a Spanish diminutive form of the Nahuatl (Aztec) word *ocotl*, meaning "torch."

Groups of these slow-growing, twenty-foot-tall perennials bloom from February to September in the Chihuahuan Desert of southern New

Fouquieria splendens, ocotillo; in flower.

Fouquieria splendens, ocotillo; stem showing spines and leaves.

Mexico on well-drained limestone slopes and rocky plains up to about 6,000 feet elevation.

Ocotillo is well prepared for survival in the land of little rain. The leaves are rarely seen, because water stress causes them to drop. After even a light rain, however, ocotillo will promptly put out leaves and then drop them when water availability falls. Each new leaf is attached to a pointed stem, or petiole, that continues to grow into a one-and-a-half-inch-long spine after the leaf has fallen (see photo). The shallow roots reach no deeper than six inches, to quickly soak up even a light rainfall. The bark contains waxes and resins that can be used to soften leather.

The Seri of Mexico sucked ocotillo's sweet nectar from the bright red blooms, while the Cahuilla of Southern California steeped the flowers in water to make a refreshing beverage. The Papago collected large quantities of the blossoms, squeezed out the nectar, and allowed it to harden into candy. The Cahuilla parched and ground the seeds to produce a meal for making cakes and mush.

The Apache crushed the roots and dusted the resulting powder on skin injuries and swelling. They concocted cough medicine from the flowers.

The Seri and Papago pierced the ears of both sexes with the sharp spines, and the Seri made necklaces of the buds, flowers, and dried pods.

The Hualapai of Arizona used the roots in a footbath. The Papago used the long stems in ceremonies to represent clouds and mountains.

In the desert, where trees are scarce, the strong, flexible stems were valuable for constructing a framework for adobe structures and brush houses. The dead-looking stems often take root, and when this occurred in these brush houses, it sometimes resulted in the curious phenomenon of ocotillo growing in a rectangular pattern out in the middle of the desert. Ocotillo canes can be planted very close together to create a living fence, in the same way as cholla, *Cylindropuntia imbricata*, p. 228.

Birds relish the seeds as a food source. The flowers have nectaries that attract a plethora of insects and hummingbirds. The willowy canes provide ideal perches for many birds to court, hunt, and spy on their neighbors.

ONAGRACEAE (Evening Primrose Family)

Gaura coccinea (syn. *Oenothera suffrutescens*)

SCARLET GAURA, BEE BLOSSOM, WAVING BUTTERFLY

Gaura coccinea (meaning "proud" and "scarlet") has every reason to be proud of itself, with its handsome little flowers that look like butterflies waving their wings. "Scarlet" is somewhat of an exaggeration, though, since the flower color transitions from white to pink and finally to red.

Like other members of the evening primrose family, *Gaura*'s flower parts are in multiples of four. The four-part stigma is in the shape of a

Gaura mollis, lizard-tail; flowers.

Gaura mollis, lizard-tail; seeds.

stubby X. The plant flowers from April to September and forms short, little fruits that are fusiform—narrow on both ends and fat in the middle. It is often difficult to say definitively to which species a given population of *Gaura* belongs because they hybridize so easily.

These perennials are usually less than two feet tall and grow in a range of soil types at elevations from 3,500 to 8,000 feet. *Gaura coccinea* is found throughout New Mexico.

As an easy-care landscape member, scarlet gaura is commercially available and will spread on its own by rhizomes. *Gaura* is popular with a wide variety of insects, although its primary pollinators are moths that arrive in the evening when the flowers open.

The Navajo include scarlet gaura as a Life Medicine and recommend it for internal injuries. They value it for treating stomachache, healing burns, and reducing inflammation of any sort. Several tribes used *Gaura* as a remedy for rheumatism, and the Kayenta Navajo cooked the plant to poultice a mother's sore breasts.

The Lakota used *Gaura* as a "horse catcher," rubbing it on their hands to help them round up their horses.

A gigantic relative, *G. mollis* (see photo), also known as *Oenothera curtiflora,* reaches five to six feet tall and has a long, undulating flower spike. It is found in similar habitats as scarlet gaura and has many of the same uses.

For information about other members of the evening primrose family, see pp. 111, 113, and 328.

OROBANCHACEAE (Broomrape Family)

(formerly in Scrophulariaceae)

Castilleja integra

WHOLELEAF INDIAN PAINTBRUSH, SQUAWFEATHER

The genus *Castilleja* was named for Domingo Castillejo, an eighteenth-century professor of botany in Spain. The species name, *integra*, tells us that the leaves are entire, not lobed or incised.

These foot-tall perennials are found in dry soil throughout New Mexico from 4,500 to 10,000 feet elevation, blooming from March to September. The small, greenish, tubular flowers are found at the top of the stems, but they are scarcely visible next to the bright colors of the bracts, or leaflike structures that masquerade as flowers.

One of the most consistent characteristics of paintbrushes is that their morphology is very inconsistent. They may be hairy or not, or sticky or not; the leaves may be lobed or not; and the bracts have many color variations.

Castilleja species are root hemiparasites that are able not only to mooch water, carbohydrates, and minerals from the host but also to take up through their haustoria various alkaloid compounds from the host roots. *Castilleja* transfers a portion of the alkaloids from the host, without chemically modifying them, to all its own parts, including the leaves, flowers, nectar, and seeds.

Native Americans have made limited use of paintbrushes as food, since the plant can accumulate toxic heavy metals from the soil as well as alkaloids from the plants they are parasitizing. However, the chemical composition of paintbrushes made them useful to the Jemez, who mixed the dried plants with stored chili seeds to prevent spoilage. The Tewa and

Castilleja linariifolia, narrow leaf paintbrush.

Hopi ate raw paintbrush blossoms, and the Tesuque sucked the nectar from the blooms.

Paintbrush tea has diuretic properties, effectively flushing the urinary tract and remedying symptoms of infections, including those of venereal diseases. The entire plant may be added to hot bathwater, providing a soothing soak for body aches—and possibly a temporary new skin color. Pregnant women drank a decoction from boiled leaves to keep their babies small for an easier delivery.

Another red-orange, but taller, cousin, *Castilleja linariifolia* (shown here) is found in northwest and central New Mexico. Hopi and Tewa women who wished to limit their family size drank a tea made from *C. linariifolia* once a month, and women and girls also used it for menstrual difficulties.

Paintbrushes are a lovely feature in a xeriscape or natural landscape, but they are difficult to establish unless the right host is provided for their roots to cozy up to. When started from seed, they may take three to four years to take off.

Paintbrushes also attract butterflies and other six-legged creatures. *Castilleja integra* is a host for caterpillars of the fulvia checkerspot butterfly (*Chlosyne fulvia*) as well as a nectar source for the adults.

For information about another paintbrush species, see p. 246.

PLANTAGINACEAE (Plantain Family)

(formerly in Scrophulariaceae)

Penstemon barbatus var. *torreyi*

SCARLET PENSTEMON, SCARLET BUGLER, HUMMINGBIRD FLOWER

Penstemons have five stamens, which is the literal interpretation of the genus name. *Penstemon barbatus* often has hairs in the throat of the flower (*barbatus* means "bearded"); in the northern part of its range, however, populations have few throat hairs. These nonhairy types are distinguished as the variety *torreyi*.

Penstemon barbatus is found in the ponderosa pine and spruce-fir zones from 4,000 to 10,000 feet elevation. The flowers appear from June through October, when its hummingbird pollinators are migrating through the Rocky Mountains to Central America for the winter. The association of hummingbirds with this plant is so strong that the Navajo name means "hummingbird food."

The Navajo made a sweet-flavored drink from the beautiful red blooms, but as a group penstemons contain too many pharmacologically active compounds for safe consumption as food.

The Navajo valued *P. barbatus* as one of their Life Medicines and used it to treat ailments such as menstrual pain, coughs, burns, and upset stomach. An extract from the root was used to treat wounds from arrows or bullets, an effective practice since the plant contains chemicals that reduce inflammation and promote new tissue growth. This extract was mixed with beeswax to soothe and heal chapped lips. Pregnant Ramah Navajo women who feared having a large child would suck nectar from scarlet penstemon flowers to keep the baby small. The Zuñi believed this plant would give them good luck in hunting, and they rubbed their rabbit sticks with the chewed root to ensure success.

Scarlet penstemon is outstanding in a landscape setting and does best in dry, well-drained soil. The plant is a vigorous seed producer, which helps maintain its presence without continually replanting. *Penstemon barbatus* and other *Penstemon* species host larvae of the dotted checkerspot butterfly (*Poladryas minuta*).

For information about other *Penstemon* species, see pp. 169 and 248.

POLEMONIACEAE (Phlox Family)

Ipomopsis aggregata (also known as *Gilia aggregata*)

SCARLET GILIA, SKYROCKET GILIA, RUBY HONEYSUCKLE, HUMMINGBIRD FLOWER

The genus *Ipomopsis* was named for the resemblance of its flowers to those of *Ipomoea*, in the morning glory family. The species name, *aggregata* (meaning "clustered"), refers to the grouping of the flowers in a spike.

Scarlet gilia is a three-foot-tall biennial with basal leaves that have sticky, stinky glands that exude "eau de skunk." Two-thirds of its height is occupied by trumpet-shaped red flowers that bloom from May through October. The plants grow throughout most of New Mexico in sandy or loamy well-drained soil in moist riparian communities or in desert scrub to pinyon-juniper and ponderosa pine communities from 5,000 to 9,000 feet elevation.

The Kayenta Navajo took the plant as an emetic, to empty the stomach prior to ceremonies, and used it as a treatment for spider bites. The Shoshone boiled it to make a decoction that was taken internally, as well as washed over the body, to treat venereal diseases. They used the same decoction as a disinfecting wash for itchy skin. For rheumatism, the Shoshone applied a warm poultice of crushed gilia to the painful area. Hopi women sometimes took a boiled leaf mixture after giving birth.

To ensure good luck on the hunt, the Ramah Navajo gave a cold infusion of scarlet gilia to hunters and then anointed their weapons with it. The Hopi sought the same good fortune by eating the dried flowers prior to hunting antelope.

Scarlet gilia is gorgeous in a wildflower garden or landscape. Because it is a biennial, it takes two years to get a stand that will continually reseed. It attracts not only hummingbirds but also insects, including butterflies, wasps, and bees.

Scarlet gilia's flowers are ornithophilous, meaning they are designed for hummingbird pollination. The red or orange tubular flowers are odorless and produce copious nectar. Migrating rufous hummingbirds (*Selasphorus rufus*) feed on scarlet gilia on their way to Central America for the winter. The calliope hummingbird (*Stellula calliope*) is another visitor.

For information about another *Ipomopsis* species, see p. 171.

Orange Flowers

ASTERACEAE (Sunflower Family)

Dyssodia papposa (syn. *Tagetes papposa*)

DOGWEED, FETID MARIGOLD

Many people find the smell of dogweed to be "disagreeable," which is in fact what *Dyssodia* means. The volatile oils responsible for the scent are found in small, reddish-brown glands on the phyllaries, peduncles, and stems. The Latin species name, *papposa*, refers to the papery bristles, or pappi, that are attached to the top of the achenes.

Dogweed is found throughout much of New Mexico, growing as an annual or short-lived perennial in disturbed areas or roadsides as well as in grasslands up to 6,500 feet elevation. The plant flowers from summer through fall.

One of the identified compounds in dogweed, α-terthienyl, is a phototoxin that causes dermatitis but also has insect-repellent properties. Another of its constituents is camphene, which gives pungency to turpentine, camphor, and essential oils such as citronella and ginger.

In spite of the unusual fragrance, some tribes used the plant for food. The Apache dried and stored the seeds, which they ground to make flour for bread. They also roasted the seeds for combining with other foods. They harvested the fresh top growth of the plants to cook as a (definitely zingy) vegetable by itself or with meat.

Extract of dogweed has been shown to limit contractions in the alimentary tract, producing a calming effect. It also has antimicrobial properties. These and other medicinal attributes were utilized by several Native American tribes. The Acoma and Laguna used the leaves to make a tea or a body rub to treat fever. Other tribes pulverized the leaves and breathed the vapors to treat headache. Dogweed was a "red ant medicine" among the Navajo, who treated ant bites by chewing the leaves and applying them as a poultice. They also used dogweed to treat illnesses, such as urinary tract problems, that produced symptoms of burning and stinging as if they had swallowed red ants.

The Tarahumara of northwest Mexico tossed crushed dogweed leaves into the water to stun fish trapped in pools, which made them easy to catch.

Dogweed's deeply divided leaves and perky yellow flowers give it potential as an attractive border plant in a xeriscape. It is a larval host for the dainty sulphur butterfly (*Nathalis iole*).

CACTACEAE (Cactus Family)

Ferocactus wislizeni (syn. *Echinocactus wislizeni*)

FISHHOOK CACTUS, COMPASS BARREL, CANDY BARRELCACTUS

Fishhook cactus is described as "fierce" (*ferus*) because of its heavy spines. The species name honors Friedrich Wislizenus, a nineteenth-century German physician and naturalist who immigrated to the United States, where he joined plant-collecting expeditions until he "retired" to the life of a physician during middle age. The Yuma and other tribes heated and bent the long spines to serve as fishhooks, giving rise to the plant's common name.

This species, along with the saguaro, is the stereotypical Southwest cactus. The plant is normally one to four feet tall and up to two feet in

Ferocactus wislizeni, fishhook cactus; fruit.

Ferocactus wislizeni, fishhook cactus; note hooked spines.

diameter, although very old plants can reach ten feet and live to be 50 to 130 years old. Larger plants tend to lean toward the southeast, earning it the name "compass barrel." This cactus grows in southern New Mexico, in sandy, rocky soil in deserts and grasslands up to 4,500 feet elevation. The yellow to orange flowers appear in August or September, and the cylindrical yellow fruits that form after pollination by cactus bees are like miniature ripe pineapples (see photo).

There are three types of spines in each areole: the long, curved "Papa Bear" spine is in the center, the "Mama Bear" spines are three to four in number and medium in diameter, and the "Baby Bear" spines are ten to fifteen in number and smaller in diameter than the others. Like other cacti, the fishhook cactus's root system is shallow, only about eight inches deep, giving it greater water-harvesting efficiency in the brief, intense rains typical of the desert.

The Seri and Apache drank the liquid from the pulp, but only in cases of emergency because the juice is apt to bring on diarrhea. The Apache parched and ground the small black seeds and then cooked them as mush. During the month of May, they ate the pulp as a green vegetable. The Pima of Arizona also drank the juice, and they still make a sweet dish by slicing the cactus and cooking it with sugary mesquite beans.

Wildlife relish the cactus's sweet fruits, with mule deer and collared peccaries high on the list of aficionados. It is also a favorite of many birds. Cattle and rabbits join in on the feast if the prickles have been removed by fire. Each fruit contains many black seeds that are ready to germinate after a journey through an animal's digestive tract.

EPHEDRACEAE (Joint-Fir Family)

Ephedra viridis

GREEN EPHEDRA, MORMON TEA

Ephedra is not a flowering plant, but a gymnosperm, like junipers and pines. Its reproductive structures are tiny male and female cones rather than true flowers, and they are the key to distinguishing one *Ephedra* species from another. *Ephedra* is dioecious, with female and male cones housed on separate plants. Photosynthesis is carried out in the slender, jointed green stems.

Most *Ephedra* species are quite similar in appearance and grow in similar habitats, in dry, rocky, open areas on slopes and foothills at 3,000 to 7,500 feet elevation.

Ephedra torreyana, joint-fir; whole plant view at White Sands National Monument.

Some Native Americans roasted and ground *Ephedra* seeds for flour to make cakes and bread. They also found that they could assuage thirst by chewing on the stems. The presence of *Ephedra* in coprolites at archaeological sites shows that it has been on the menu since ancient times.

Ephedra can be brewed into a refreshing and soothing beverage, as the Mormons discovered in their trek across the West. The Chinese have used *Ephedra* for thousands of years as a treatment for all sorts of respiratory problems; the beneficial effect is believed to be due to the action of ephedrine and pseudoephedrine alkaloids. Native Americans have employed the plant for the same purposes. Curiously, several analyses of the New World species of *Ephedra* indicate that they do not contain ephedrine and have only negligible amounts of pseudoephedrine. However, both Old World and New World species do accumulate several nitrogenous compounds that have known pharmacological activity, including quinoline-2-carboxylic acids in the stems and cyclopropyl amino acids in the stems and roots. These compounds are probably the basis for the widespread use of *Ephedra* in both the Old and New Worlds as a traditional medicine.

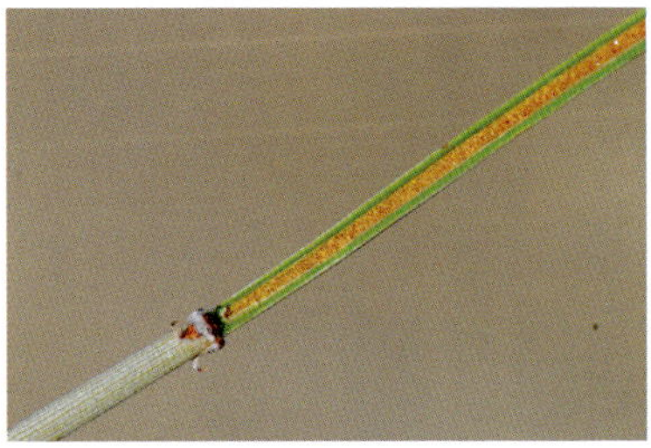

Ephedra spp.; split stem showing orange pith as evidence of optimum concentration of active ingredients.

When the active ingredients have reached optimum concentration in the stems, the pith becomes bright orange-yellow, as the photo depicts.

The Acoma and Laguna made use of its diuretic and antiseptic properties in treating urinary tract infections, and the Zuñi and Tewa drank it to soothe the symptoms of venereal diseases, which often involve pain in the urinary tract. Some *curanderas* (traditional natural healers) recommend *Ephedra* for liver problems.

Ephedra torreyana (see photo on p. 27) is found throughout New Mexico. Its male cones look like fragile pinecones with lots of little "gecko feet" sticking out (see photo). Its female cones are yellow-green, papery, and tightly closed.

Ephedra trifurca (see photo) is located in southern New Mexico. The distinguishing features are its translucent pink-beige papery female cones that open wide at maturity and its long, thin leaves in sets of three that resemble a fork (*trifurca*).

Ephedra viridis is found in northwest and central New Mexico. The female cones are brown and look like Kermit the Frog's Muppet mouth. *Ephedra viridis* provides winter forage for both large and small mammals

Ephedra torreyana, joint-fir; male cones.

Ephedra trifurca, Mormon tea; mature female cones.

but is toxic to pregnant sheep and cows. It reproduces easily from seed, from stem cuttings, and from sprouting at the crown, making it useful in landscaping and restoration projects. The Navajo used *E. viridis* for weight loss and for a stimulant tea.

EQUISETACEAE (Horsetail Family)

Equisetum laevigatum

SMOOTH HORSETAIL, SCOURING RUSH

Equisetum is not a flowering plant, but a gymnosperm that is related to ferns, such as lip fern (*Cheilanthes*) and bracken fern (*Pteridium*), found in New Mexico. Its reproductive structures are tiny male and female spore-bearing structures collected into conelike strobili rather than true flowers.

Equisetum laevigatum translates from Latin as "smooth horse tail." This herbaceous perennial species has upright, hollow, jointed stems that can reach five feet tall, with underground rhizomes. It prefers the edges of lakes and streams in shady areas with moist soil. It has the potential to become invasive if left unmanaged.

Equisetum laevigatum, smooth horsetail; note hollow, ridged stems.

Although *E. laevigatum* is known to be a neurotoxin for horses, the Tewa are said to have used horsetail as food for their horses. Some Native Americans ground the plant with cornmeal to make a ceremonial bread. In the ruins of Antelope House at Canyon de Chelly, archaeologists have found evidence that the Ancestral Puebloans used *Equisetum* for food many centuries ago.

The Kayenta Navajo prepared an infusion of horsetail and used it as a lotion to rub on aching backs. Horsetail served as horse medicine for the Cheyenne and Blackfoot. Modern herbalists make healing claims for *Equisetum*, touting its astringent, diuretic, and hemostatic properties. Extracts are said to help with urinary and prostate problems, to help repair damaged lung tissue, to alleviate edema, and to reduce the pain of rheumatism.

The stems contain silica and were used to polish bone and wood items, as well as pots and pans. The hollow stems made handy straws for drinking or dispensing medicine.

FABACEAE (Bean Family)

Ottleya wrightii (syn. *Lotus wrightii*)

WRIGHT'S DEERVETCH, RED AND YELLOW DEERVETCH, WIRY LOTUS

Deervetch's genus name honors Alice Ottley, a twentieth-century California botanist and author. The species name recognizes Charles Wright, a nineteenth-century American botanical explorer who collected plants in the Southwest.

This perennial, with its multibranched stems and compound leaves, has an open, airy look. The orange to yellow flowers, which bloom from April to August, are reminiscent of the old-fashioned candy corn. Deervetch grows in the western half of New Mexico in dry, rocky, or sandy areas up to 5,500 feet elevation, mainly in the pinyon-juniper zone.

One of the Life Medicines of the Navajo, deervetch acts as a cathartic to clear out the digestive tract. The Ramah Navajo boiled the leaves for a drink to settle the stomach (early Alka-Seltzer!). The Zuñi chewed the roots to make a poultice for swelling believed to be caused by bewitchment by a bull snake.

Deervetch is attractive in natural landscape settings as well as in xeriscapes, where it performs best with very little supplemental water. Birds and small mammals eat the seeds.

LINACEAE (Flax Family)

Linum puberulum (syn. *Mesynium puberulum*)

PLAINS FLAX

Flax has been known since ancient times as *linon* in Greece. *Puberulum* is a Latin word meaning "with tiny hairs," referencing the minute hairs that cover the stems and sepals of this plant.

Plains flax grows throughout New Mexico in sandy or gravelly soil in the desert and pinyon-pine zone up to about 7,000 feet elevation. This delicate plant is easily overlooked because its color scheme is well camouflaged in the surrounding plants and soil. The flowers have a reddish "eye" in the center, and the leaves are narrow and linear.

Plains flax is counted among the Life Medicines of the Navajo, who boil down the leaves to make a decoction to ease heartburn. The Zuñi crushed the fruits to obtain a juice for treating eye inflammation. The Acoma and Laguna used the copper-yellow blooms to produce yellow paint.

Linum aristatum, bristle flax.

Another orangish-yellow-flowered flax, but without a distinguishing "eye," is *Linum aristatum*, or bristle flax (see photo), which is typically found in dry sandy soil in southern New Mexico.

For more information about flax, see p. 167.

MALVACEAE (Mallow Family)

Sphaeralcea angustifolia

COPPER GLOBEMALLOW, NARROW-LEAF GLOBEMALLOW

The genus name, *Sphaeralcea*, comes from the Greek *sphaira* (globe), describing the spherical seedpods, and *alcea*, the name given by Dioscorides to mallow plants. Globemallows are covered with stellate (star-shaped) hairs, giving them a grayish look. These perennials often grow in sandy or caliche soil, usually from 5,000 to 8,000 feet elevation. The flowering season is generally May through October. The fruits resemble miniature cheese wheels, with sections that split open to release the seeds.

Sphaeralcea angustifolia, copper globemallow; field view.

Sphaeralcea coccinea, caliche globemallow; close-up of flower, and leaf with palmate venation.

Sphaeralcea coccinea, caliche globemallow; seedpod.

For the Navajo, globemallow is a Life Medicine, useful for treating colds and headache, stimulating the appetite, and settling an upset stomach. Native Americans used the roots to help make casts for broken bones, to stanch bleeding, to soothe teething pain, and to poultice sores and snakebites.

Globemallows contain mucilage that soothes or stiffens, depending on the application. The Navajo name "gummy medicine" acknowledges this sticky property. After washing their hair, some Native Americans would rinse it with an infusion of globemallow as a conditioner to add shine and softness. If they desired a stiffer "hairdo," they would allow the globemallow infusion to dry in the hair without rinsing it out.

Globemallow was also used in maintaining the home. A paste from the pulverized roots was used to harden adobe floors in pueblos. Today the Santo Domingo still add boiled globemallow to house paint to improve its quality.

All *Sphaeralcea* species are attractive elements in natural landscapes and valuable as part of native seed mixtures for rehabilitating roadsides and other disturbed areas.

Distinguishing one species from another is based largely on leaf shape. The leaves of all the species described here are palmately veined but differ in the extent to which the outer edge of the leaf is lobed around the venation pattern.

Sphaeralcea angustifolia grows throughout New Mexico and has narrow leaves that are longer than those of other species.

Sphaeralcea coccinea (caliche globemallow, scarlet globemallow; see photo) grows throughout New Mexico and has leaves that are very palmately divided. It is the most economically and ecologically important species and the shortest of all the *Sphaeralcea*, but with a far wider range throughout western North America than its colleagues. It is a dominant species in the Great Plains short-grass and mixed-grass prairie

Sphaeralcea incana, gray globemallow; note mitten-shaped leaves with two "thumbs."

Sphaeralcea digitata, finger globemallow; note very narrow leaves.

Sphaeralcea parvifolia, small-leaf globemallow; note rounded leaves with palmate venation.

communities and is also common in desert grassland, pinyon-juniper, and sagebrush communities. It has a deep taproot as well as rhizomes, by which it colonizes the surrounding area. It is a food source for livestock and wildlife, including birds and large and small mammals.

Sphaeralcea digitata (finger globemallow, juniper globemallow, bird's-foot mallow; see photo) is found in central and western New Mexico. It has small, palmately divided leaves that look like tiny fingers, with lobes just a few millimeters wide.

Sphaeralcea incana (gray globemallow, soft globemallow; see photo) is found in central and western New Mexico and has leaves shaped like a mitten with a thumb on each side.

Sphaeralcea parvifolia (small-leaf globemallow, smallflower globemallow; see photo) is found primarily in the Four Corners area and has rounded leaves.

Sphaeralcea species are host to the caterpillars of several butterflies, including painted lady (*Vanessa cardui*), west coast lady (*Vanessa annabella*), small checkered-skipper (*Pyrgus scriptura*), and northern white-skipper (*Heliopetes ericetorum*).

MONTIACEAE (Miner's Lettuce Family)

(sometimes placed in Portulacaceae)

Phemeranthus aurantiacus (syn. *Talinum aurantiacum*)

ORANGE FAMEFLOWER, TALINUM

The flowers of these little drops of sunshine are ephemeral, opening in the afternoon for only two to four hours. The noble-sounding genus name, *Phemeranthus*, translates directly as "fameflower"; it was given in honor of Pheme, the gossip-mongering Greek goddess of fame and renown. The suffix *anthus* means "flower." The species name, *aurantiacus*, refers to the rich orange-red color of the flowers.

Orange fameflower is a petite perennial, scarcely achieving four inches in height, and is easy to overlook except when in full bloom. Plants grow in southern and eastern New Mexico, in rocky soil in desert scrub and grassland communities from 2,800 to 6,500 feet elevation. The flowers bloom from April to November. The leaves are succulent and have a groove along the center.

The large taproot is edible and was once on Native Americans' menu. Wildlife browse the foliage.

PAPAVERACEAE (Poppy Family)

Eschscholzia californica subsp. *mexicana*

MEXICAN POPPY, MEXICAN GOLD POPPY

The state flower of California, *Eschscholzia californica*, was named in honor of Johann von Eschscholtz, a nineteenth-century German surgeon, professor, and zoologist who first collected this plant while exploring near San Francisco.

Eschscholzia californica subsp. *mexicana*, Mexican poppy; field view.

This subspecies is found in southwest New Mexico, where it grows in well-drained soil in desert grasslands and hillsides from 4,500 to 6,500 feet elevation. The annual plants are usually less than a foot tall and bear large, brilliant flowers amid fernlike foliage. The flowers bloom from February to May in colors ranging from orange to white or pinkish. The upright pods split to release a multitude of seeds, which germinate easily nearby, eventually covering large areas with a blaze of color.

There are many medical applications for Mexican poppy, ranging from a sedative for crying babies to a facial cosmetic made of the pollen. The flowers were boiled to prepare a decoction used to kill head lice. The roots were used as a remedy for venereal disease and to relieve toothache. The crushed seedpods were applied to a mother's breasts when she was ready to wean her baby.

Gophers eat the roots, but there is little information about the plant's value to other wildlife. Not surprisingly, Mexican poppy is an enduring favorite in gardens and landscapes. Seed of several varieties is commercially available. It is often included in native flower mixes for use in land reclamation projects such as roadside rehabilitation.

PINACEAE (Pine Family)

Pinus edulis

PINYON PINE, TWO-NEEDLE PINYON, NUT PINE

Pinyon pine, the state tree of New Mexico, has long been a valuable food source for both humans and wildlife, as reflected in its name, *Pinus edulis,* meaning "edible pine."

Pinyon pine is found throughout New Mexico at 4,000 to 9,000 feet elevation, often associated with juniper species below 7,000 feet and with ponderosa pine above 8,000 feet. It is very slow growing and drought

Pinus edulis, pinyon pine; field view.

resistant. When it reaches twenty-five years old, it begins to produce cones, but significant production does not begin until the tree reaches seventy-five years, and even then good production occurs only every five years or so. Trees may grow to fifty feet tall, and larger trees in a stand may be up to four hundred years old.

Pinus edulis, pinyon pine; staminate (male) cones.

A distinguishing feature of pinyon pine is the pair of one- to two-inch-long needles in a bundle. The rough, scaly bark often has resin exuding from it. Pine trees are gymnosperms, reproducing by male and female cones rather than flowers. Pinyon pine is monoecious, with female and male cones borne on separate parts of the tree; the female cones are solitary and look like a small, typical pinecone, while the male cones are small, orange, and clustered in large groups. Reproduction requires up to two years before the edible nuts are mature.

Pinyon nuts are an important commercial crop in the United States today, with the market estimated to be about $10 million for an average annual harvest of one million pounds. The nuts are a valuable source of high-quality protein and oil, containing 14 percent protein, 62 to 71 percent oil, and 18 percent carbohydrates, which compares favorably with the nutrient content of other nuts, such as pecans and walnuts. Pinyon protein is "complete," meaning it contains all twenty amino acids needed for human nutrition.

A pottery glaze is made from resinous sap of pinyon pine.

The value of pinyon nuts today pales in comparison to their importance in the past. Packing almost three thousand calories per pound of nuts, they provided a vital buffer against famine. Shells have been found virtually everywhere Ancestral Puebloans lived. At Salmon Ruins, near the Four Corners area, evidence points to pinyon nuts being as essential as squash in ancient people's diet.

The nuts are gathered in late summer or fall and are eaten roasted, raw, or ground into flour for baking. Because many animals compete with humans for this valuable food source, some groups learned early on how to outwit their competitors by gathering the cones before they opened, then roasting them to release the seeds. Evidence for this practice was found at Salt Bush Ruin at Bandelier National Monument, where many whole cones were found in storage.

Native Americans chewed the tree sap, or pitch, like chewing gum to clear congestion or burned it to produce a vapor to open the breathing passages. The sap, sometimes mixed with clay, also was used as a liniment on sores and cuts. It was boiled in water and drunk to eliminate parasites, including worms.

The resinous sap was also a waterproof sealant for pottery (see photo) and baskets and a glue for attaching feathers to arrow shafts, sinews to bows, and turquoise to silver settings. The soles of moccasins were sealed with pinyon pitch to keep them flexible and waterproof.

The Jemez still boil the resin to extract a lovely blue-green paint. By mixing the resin of old and young pinyon trees, red paint may also be obtained. Mixing the pitch with sumac leaves and yellow clay makes a black dye for wool.

Birds, bears, small mammals, pronghorn, and deer use this plant as a major food source. Pinyon pine is host to larvae of the pine white butterfly (*Neophasia menapia*), whose adults feed on the nectar of rabbitbrush and *Monarda* species (see pp. 51 and 202).

TYPHACEAE (Cattail Family)

Typha latifolia

CATTAIL, TULE

Desert dwellers have a keen appreciation for permanent sources of water and the useful plants, such as cattails, associated with them. *Typha*, a common water plant around the world, has been called by its Greek name since the time of Theophrastus more than two thousand years ago, when its seed fluff was used for stuffing beds. *Latifolia*, meaning "broad leaves," recognizes the strong, flexible, six-foot-tall leaves that rise from the rhizomes anchoring the plants in wet or muddy areas. Even if cattails are not surrounded by water, their very presence says that water is near the surface on a permanent basis. Cattail colonies are usually found below 7,000 feet elevation.

Flowering takes place from May to July. The flowers look like two fat brown cigars impaled one above the other on a shish kebab stick. The male flowers, the top "cigar," shed prodigious quantities of pollen until only the stick is left. The female fruits seem to be almost entirely fluff, with all manner of womanly uses. Women collected the absorbent material and kept it on hand for their monthly periods and for packing around babies in their cradles. As the fluff became wet, it was removed and replaced with fresh material.

All parts of cattails are edible, providing a rich source of carbohydrates. The thick, wax-covered rhizomes contain up to 50 percent carbohydrates and can be baked like a potato, with much the same texture and flavor. Many modern Native Americans eat the rhizomes by themselves or cook them with meat. The young shoots have a fresh flavor reminiscent of cucumber. They are eaten raw with a bit of salt, made into pickles, or mixed with cornmeal prior to baking. Steamed young shoots taste like cabbage. As the stems lengthen during the growing season, the sugary sap becomes another food source; children enjoy chewing the stems like sugarcane. As the flower spikes begin to form, they can be harvested and boiled like okra in soup.

Many tribes collect the protein-rich pollen just as it matures and use it like flour to make pollen cakes, which have a slightly sweet flavor. In the past, mature seeds were harvested, ground, and formed into cakes that were roasted over hot coals. Archaeologists have found much cattail pollen in prehistoric shelters and in desiccated human feces.

In addition to being a dependable and nutritious source of food, cattails were also used for medicine. The young flower heads were eaten as a remedy for diarrhea, and the roots were made into an ointment for burns and wounds. The seed fluff was used like powder to prevent chafing.

Since cattails grow in or very near water, they have come to symbolize water in ceremonies. The Acoma and Laguna shake the ripened cattails in rain dances to encourage the formation of clouds. The Navajo use the plant in the Lightning Way Ceremony and hang cattail mats in their hogans to protect them from lightning. The Mescalero Apache use the long leaves to cover the ground in the ceremonial tepees used during puberty rituals. The Navajo make necklaces and wristbands from the leaves for the Male Shooting Chant.

The long, straplike leaves are perfect for making all sorts of mats and baskets. Groups such as the Isleta and Zuñi use the stems as lathing over their large, weight-bearing vigas (roof beams). The cattail thatch supports the mud or adobe covering the juniper or pine beams.

Many wildlife species feed on cattails, including geese, elk, and muskrats, while a great variety of waterfowl and shorebirds use it for cover.

Yellow Flowers

APOCYNACEAE (Dogbane Family) (formerly in Asclepiadaceae)

Asclepias latifolia

BROADLEAF MILKWEED

Both the common name and the Latin name, *latifolia* (broad leaves), are descriptive of the three- to four-inch-long leaves, which are two inches wide and feature prominent veins that any phlebotomist would love. The

Asclepias latifolia, broadleaf milkweed; field view.

Asclepias latifolia, broadleaf milkweed; seedpod.

flower clusters emerge from the axils where the leaves attach to the plant, with individual flowers scrumptiously suggestive of buttery kernels of corn.

Broadleaf milkweed is a perennial, growing in a variety of soil types in grassland, desert, and pinyon-juniper zones up to 7,000 feet elevation. It blooms from June to September and is found throughout New Mexico, except in the northwest.

In spite of its tasty appearance, broadleaf milkweed contains a portfolio of cardiac glycosides that at the wrong dosage are not conducive to longevity. In judicious amounts, however, broadleaf milkweed can be therapeutic instead of lethal. The Isleta people dried the leaves and stems and ground them into a powder to use as snuff for respiratory congestion. They used the sap to treat lung ailments, including asthma. They ground the roots into a powder that was used as a strong intestinal cleanser and to treat joint pain, expel worms, and remove warts.

Women among several Native American peoples used *Asclepias* roots for contraception. When this treatment failed, the roots' effects on smooth muscles were useful during and after childbirth.

For information about other milkweed species, see pp. 215 and 266.

ASTERACEAE (Sunflower Family)

Baileya multiradiata

DESERT MARIGOLD

Desert marigold's genus name honors Jacob Whitman Bailey, an early-nineteenth-century pioneer in the use of microscopy, whose son became a botanist. The species name, *multiradiata*, is based on the "many ray flowers" in the flower head.

This plant's fuzzy, gray-green basal leaves and long-lived, rich lemon-yellow flowers with toothed petal tips make it an attractive xeriscape feature, whether in desert communities, where it typically grows from

Baileya multiradiata, desert marigold; field view.

2,800 to 5,000 feet elevation, or in a cultivated setting, where it needs no pampering and easily reseeds. Whether in the wild or in a garden, desert marigold needs well-drained sandy or gravelly soil. It grows as a biennial or sometimes a short-lived perennial in southwest and south-central New Mexico.

The Acoma and Laguna rubbed the plant under their arms as a deodorant. The Jemez mixed it with clay to make adobe buildings and plaster.

Desert marigold is toxic, both fresh and dry, especially to sheep, which love the flowers and seed heads but develop severe weakness and a rapid heartbeat after consuming them. Several days of this malaise are usually followed by complete recovery. An adult sheep would need to consume as much as sixty pounds of the plant to find itself pushing up the next crop of daisies. Rabbits and rodents do not seem to be disturbed by the poisons.

Desert marigold seeds attract birds such as doves, finches, and quail, as well as small mammals.

ASTERACEAE (Sunflower Family)

Berlandiera lyrata

CHOCOLATE DAISY, LYRELEAF GREENEYES

Jean-Louis Berlandier, a French botanist who trained in Switzerland, one of the motherlands of chocolate delicacies, appropriately gave his name to chocolate daisy, which exudes all the wonderful aromas traditionally associated with the world's finest "fruit." The leaves are somewhat in the shape of a lyre, a musical instrument that has a wide bottom and narrower top.

Berlandiera lyrata, chocolate daisy; field view; note "green eyes."

Berlandiera lyrata, chocolate daisy; immature seed head.

Chocolate daisy is no lazybones. Its flowers open very early in the day to regale morning walkers with their divine aroma and bright, cheery color. By early afternoon, as the day heats up, chocolate daisy closes its ray flowers and assumes its green-eyed look, created by the smoothly curved phyllaries cupped around the velvety green disk flowers. Immature flower heads (buds) also look like "green eyes."

Chocolate daisy is found over most of New Mexico, except for the northwest. Plants grow in grasslands to juniper-oak woodlands from 2,800 to 7,000 feet elevation. The yellow ray flowers, which have maroon-colored veins on the undersurface, are females, while the green disk flowers have fertile male parts and infertile female parts. The seeds are borne only by the ray flowers; the tuft sticking up in the middle of the seed head at maturity is just seedless flower parts.

The Acoma and Laguna burned the dried roots and inhaled the smoke to control their nervousness and gain courage. They also mixed the flowers with meat as a seasoning.

This perennial plant, with its attractive flowers and foliage, is popular in xeriscape gardens. It blooms the entire growing season and is content in dry, rocky limestone soil. Trying to grow it in wet, heavy soil will result in root rot. Chocolate daisy is highly deer resistant.

ASTERACEAE (Sunflower Family)

Centaurea melitensis

TOCALOTE, MALTA STAR THISTLE

According to Greek mythology, this plant was used to heal the wounded foot of Chiron, one of the centaurs of Thessaly. The species name, *melitensis*, as well as one of the common names, indicates that the plant's homeland was Malta.

Like some of its thistle cousins, tocalote has a disreputable character arising from its general spininess and toxic chemistry. The spiny bracts surrounding the fairly inconspicuous yellow flowers give it a starlike appearance. The stems and branches have wings and are covered with dense white hairs.

From its origin in the Mediterranean region, tocalote was introduced in California as a stowaway in the late 1700s and has spread through several states including New Mexico. It is classified as a Class B noxious weed, meaning that while it is very abundant in some regions, it has not reached others, and the goal is to contain its spread. The plant inhabits disturbed areas, growing from seed each spring in agricultural areas, along roadsides, and in chaparral and pine-oak woodlands up to about 5,000 feet elevation.

When horses have access to this fresh spring growth, they can develop brain damage and die from the sesquiterpene lactone toxins. They must ingest at least 50 percent or more of their body weight in plant material over a period of up to three months before the symptoms of "chewing disease" occur. These symptoms include continuous chewing, difficulty drinking and swallowing, and aimless walking. Sheep and other animals are not affected by tocalote's toxins.

ASTERACEAE (Sunflower Family)

Engelmannia peristenia

ENGELMANN DAISY, CUT-LEAF DAISY

Engelmann daisy is a monotypic plant, meaning it is the only species in its genus. The genus name honors the nineteenth-century German doctor and botanist George Engelmann, who came to America as a youth and is most famous for his work in describing over one hundred cactus species.

Engelmann daisy occurs in most counties of New Mexico, growing in a wide variety of soil types in desert scrub, grassland, and pinyon-juniper communities up to 6,500 feet elevation. The deep taproot gives it drought tolerance, and it can be found blooming when less well-endowed species are not to be seen. This perennial overwinters as a rosette and sends up attractive yellow flowers from March to July, or even later in some locations. The flowers are fresh and cheerful in the early morning, but by afternoon the petals curl backward (see photo), giving the plant a bedraggled look.

The ray flowers around the edge of the flower head are fertile females; the disk flowers are perfect (having both male and female structures), although the female parts are sterile. The small, dry fruits—achenes, as in all members of the sunflower family—have a most unusual appearance, resembling the convoluted cortical folds of the brain.

The leaves have a high protein content, up to 25 percent in early spring, and are very palatable to cattle, sheep, and goats, as well as to deer and rabbits. You will not find Engelmann daisy in an overgrazed pasture, because the livestock will decimate it. Small mammals and birds also feed on it and use it for cover. Hummingbirds and various insects, including butterflies and bees, enjoy the nectar.

Engelmann daisy is very striking in xeriscape plantings, but it is an "ice cream" plant for deer, and if they live in the area, the plants may soon be deer food. Seed is commercially available and is often part of a mixture of species planted on roadsides and embankments to reestablish vegetation.

ASTERACEAE (Sunflower Family)

Ericameria nauseosa (syn. *Chrysothamnus nauseosus*)

RABBITBRUSH, GOLDENBUSH, CHAMISA

The genus name *Ericameria* refers to the similarity of rabbitbrush's leaves to *Erica*, or heath. The synonymous genus name, *Chrysothamnus*, means "golden bush." Its Navajo name mirrors that visual image, meaning "big yellow on top." The species name *nauseosa* refers to "seasickness," which is notably characterized by vomiting.

Rabbitbrush's masses of golden flowers define the late summer and fall landscape in New Mexico, where this plant occurs in almost all counties in pinyon-juniper and sagebrush communities between 4,000 and 8,000 feet elevation. The stems are covered with heavy, woolly hairs and are topped with flower heads containing long, tubular disk flowers. Rabbitbrush is a prolific producer of tiny seeds adorned with long, fluffy pappi to help them set sail on the wind.

Rabbitbrush is an eco-friendly chemical manufacturing facility. It contains about 7 percent rubber by dry weight and was tested as a source of rubber during World War II. It also contains about 35 percent resin by dry

Ericameria nauseosa, rabbitbrush; gone to seed.

weight in the leaves and trichomes (hairlike glandular structures). This resin contains aromatic terpenoid compounds that may be useful for the manufacture of varnishlike plastics. Some of the terpenoid compounds have been tested as nematicides and insecticides.

These aromatic compounds were valuable components in the Indian medicine cabinet. The Navajo used rabbitbrush leaf tea to bring down fever, treat smallpox, and relieve arthritis pain. The flowers and roots were used to make a tea to relieve tuberculosis. In the ob-gyn department, the Navajo made an anodyne for menstrual pain from the roots. Other tribes used rabbitbrush to make a sanitary napkin to promote healing after childbirth.

In spite of its pungent chemistry, the Navajo cooked the green leaves together with wild onions in meat dishes. The seeds were added to cornmeal porridge and bread.

Some rabbitbrush plants are decorated with little cotton balls along the stems. These are actually galls made by small flies in the Tephritidae family. Chemicals from the developing larvae trigger the formation of tumorlike tissue that encases the pink larvae. Indian tribes used the cottony galls to treat toothache and stomach problems.

One of rabbitbrush's most important uses is as a dye for coloring wool, cotton, and baskets. Several tribes use the plant to produce a colorful palette of yellow and green dyes—yellow from the mature flowers and green from the immature flowers, leaves, and green bark.

In the past, Native Americans used rabbitbrush for an eclectic assortment of other life needs. The Hopi used it to weave a belt for weddings, made arrows from the wood, and made windbreaks with the bushes.

The Zuñi used the stems for decorating pottery and to make baskets and brooms.

Wildlife species, both large and small, utilize rabbitbrush as a reserve food. Its palatability varies by ecotype, but all ecotypes are more palatable in the fall and winter.

This deep-rooted perennial grows quickly from seed, is drought and cold tolerant, and reaches up to eight feet tall, making it useful for landscaping and for rehabilitating degraded mining sites and disturbed areas along roadsides.

Planting rabbitbrush in landscapes ensures a hearty population of birds, butterflies, and a host of other insects that partake of the plant's nectar. Larvae of butterflies such as the gray hairstreak (*Strymon melinus*) and northern checkerspot (*Chlosyne palla*) feed on it. Adult butterflies such as the pine white (*Neophasia menapia*) enjoy its nectar.

ASTERACEAE (Sunflower Family)

Gaillardia pinnatifida

YELLOW GAILLARDIA, RED DOME BLANKETFLOWER, YERBA DEL SOL

The genus *Gaillardia* was named for the well-known eighteenth-century French lawmaker Gaillard de Charentonneau, who financed botanical

Gaillardia pinnatifida, yellow gaillardia; field view.

research and exploration. The species name, *pinnatifida*, is Latin for "pinnately lobed leaves."

The flowers of this perennial are large and showy, consisting of yellow ray flowers with reddish disk flowers in the center. The globose seed heads are attractive in their own right, as seen behind the flower in the main photo. The leaves have stiff hairs and are clustered at the base of the plant. Yellow gaillardia grows in a wide variety of soils in desert, grassland, and pinyon-juniper communities from 3,500 to 7,000 feet elevation. Like other gaillardias, it is easy to grow from seed and is very attractive in natural or waterwise landscapes.

Yellow gaillardia contains alkaloids that may cause liver damage if used over an extended period of time.

The Navajo prepared a tea and a poultice of the leaves to treat gout. They also used the plant for heartburn and nausea.

Native American women believed that their fertility would be improved if they drank a tea made from yellow gaillardia for seven days in a row. The tea was also used as a diuretic and as a pain reliever for bladder infections. An old folk remedy for sinus headache involved making a poultice of the entire plant and applying it to the forehead. If that failed, a snuff made from the flowers could be used to induce sneezing to relieve a stuffy nose.

For information about other *Gaillardia* species, see p. 11.

ASTERACEAE (Sunflower Family)

Grindelia squarrosa

CURLY-CUP GUMWEED, ROSINWEED, TARWEED

Grindelia was named in honor of David Grindel, a German botany professor, pharmacologist, and physician who lived during the late eighteenth and early nineteenth centuries. The Latin species name, *squarrosa*, means "spreading horizontally or at right angles," a description of the phyllaries, which fan out with a backward curl around the lower part of the flower head. These phyllaries also give it the common name "curly-cup."

Curly-cup gumweed can be a biennial or perennial, depending on its environment. It is found throughout New Mexico in disturbed or overgrazed land in a number of communities from 4,000 to 7,500 feet elevation. The flowers open for business in late summer and fall. Most populations of curly-cup gumweed have flower heads with yellow ray flowers, but there are also large populations with only disk flowers, as shown here, that are sometimes classified as *G. nuda*.

A defining feature of this plant is the sticky white sap that exudes from glands in the flower heads, leaves, and stems. This sap contains a host of useful compounds, including tannins, aromatic oils, alkaloids, and resins. The resins are made up of diterpenoid derivatives, including the major constituent, grindelic acid, which shows antimicrobial activity. Other key compounds are flavonoids, sterols, and sapogenins. The flavonoid acacetin has potential antitumor activity and acts as a vasorelaxant, lowering blood pressure. Quercetin, another flavonoid, lowers cholesterol.

Grindelia arizonica, Arizona gumweed; note upright phyllaries.

These medicinal properties are reflected in the many uses that Native Americans had for curly-cup gumweed. They used it to relax the spasms that accompany asthma and severe coughing and to relieve congestion in the chest and head. Indian tribes in the northern Great Plains chewed the entire flower head like chewing gum. Imagine the refreshing oral and nasal sensations experienced by someone afflicted with a bad cold!

American Indians also treated skin sores and rashes in both humans and horses with curly-cup gumweed. They used the fluid extract to reduce the inflammation of poison ivy rashes, as well as the sores of smallpox and measles. The Pawnee washed saddle sores on horses' backs with the extract. Curly-cup gumweed was used as a disinfectant wash for wounds, and the sticky sap was used to hold the edges of wounds together. It was also used to induce abortions and treat liver trouble. The Navajo used an extract as a diuretic, as a treatment for bladder pain and venereal diseases, and as an emetic to induce vomiting in preparation for certain ceremonies. The Navajo dried the plant and crushed it into a powder for dusting over a freshly washed face to prevent pimples.

Yellow to gold dyes can be extracted from the flowers, stems, and leaves.

Another gumweed, *Grindelia arizonica* (see photo) has upright phyllaries. This species hybridizes with *G. squarrosa* to form intermediate populations.

ASTERACEAE (Sunflower Family)

Gutierrezia sarothrae

BROOMWEED, SNAKEWEED, MATCHBRUSH

Broomweed is said to occupy more real estate than any other perennial plant in the Four Corners area. It is also one of the most important medicinal plants in New Mexico.

The genus name honors Pedro Gutierrez, a Spanish botanist associated with the Royal Botanical Gardens in Madrid during the early 1800s. *Sarothrae* is from the Greek word *sarum*, meaning "broom," which in fact is what this plant was used for.

Broomweed is found throughout New Mexico in almost every community up to alpine elevations and is common between 3,800 and 7,500 feet. Broomweed readily colonizes disturbed areas and rapidly increases in overgrazed rangeland. It may also compete through allelopathy, a chemical warfare that discourages the germination and growth of neighboring plants.

The plant grows about knee high under favorable conditions, and the edges of its globose profile are covered with small yellow flower heads that bloom from August through the fall. The aromatic oils are found in resinous glands on the leaves, which give off a pleasant pinelike smell and make broomweed useful as kindling for fires.

The resinous glands have a portfolio of potentially valuable pharmaceutical compounds, including flavonoids and saponins, as well as a host of volatile fragrant compounds such as limonene and pinene. Native Americans have a number of uses for broomweed based on these medicinal properties.

Among American Indians broomweed is very important ceremonially for warding off evil and is a Life Medicine for the Navajo. They use broomweed to heal wounds in both humans and animals, including castration cuts in livestock. The chewed pulp relieves the sting of insect and spider bites, and the boiled plant is used to treat snakebite in sheep. The root is used to help deliver the placenta during childbirth and to treat diarrhea and urinary problems. Broomweed's effectiveness is said to rival that of aspirin for the relief of arthritis pain, for which the tops of the plant are boiled in water and the resulting tea is used both for drinking and for adding to bathwater.

Broomweed is an indicator plant for ruin mounds because it often grew around pueblos.

ASTERACEAE (Sunflower Family)

Helianthus annuus

COMMON SUNFLOWER

The "annual" (*annuus*) "sunflower" (*helios*, "sun," and *anthos*, "flower") is the ancestor of the modern commercial sunflower, the largest indigenous American oilseed crop now grown around the world. For more than four thousand years, Native Americans used wild sunflowers for both food and cooking oil and, through genetic selection, gradually improved seed size and oil content. The Spanish conquistadors took sunflower seeds

Helianthus annuus, common sunflower; close-up of flower head and bud wrapped with leaflike phyllaries.

Helianthus annuus, common sunflower; immature seeds.

back to Europe, where the plant was largely unappreciated. Not until almost four hundred years later, when the seeds reached Ukraine in the mid-1800s, did genetic improvements begin that transformed sunflowers into a commercial crop. Today sunflowers are grown to produce a premium vegetable oil for the global market.

In the wild, common sunflower grows up to ten feet tall, featuring flower heads of bright yellow ray flowers and reddish-brown disk flowers that bloom from July to fall. The leaves are large and hairy, with a rough texture like that of a cat's tongue. Sunflower is widespread in the Southwest and is found in most ecological zones up to 7,000 feet elevation.

Many Southwest Indians made extensive use of sunflower not only for food but also for personal care products, medicine, and ceremonial needs. The Navajo ground the seeds into a meal for cakes and bread, or toasted and ground the seeds to extract the oil. Many tribes ate seeds as a high-protein, high-energy supplement when they went hunting or warring, and nursing mothers ate sunflower seeds to give strength to both baby and mother. The Pima and Maricopa made chewing gum from the pith of the stems. Early pioneers made a coffee substitute from the shells, with the deception made complete by the dark color imparted by roasting.

Sunflower contains a number of active ingredients with the potential to address a variety of health issues. Its "chemical factory" produces several types of tannins and terpenes, particularly sesquiterpene lactone and diterpenes. Various Southwest tribes utilized sunflower as a poultice for snakebites, as an ointment for skin sores and inflammation, and as a treatment for kidney ailments. Sunflower tea was used to treat pulmonary problems, rheumatism, high fever, and even screwworm sores on horses. For wart removal, the Navajo burned the pith and placed it on the wart.

The Navajo used the hollow stem to make a flute with four notes to mark the beat for grinding corn in the Enemy Way Ceremony. San Ildefonso women decorated their dresses with sunflowers for the Corn Dance. The Zuñi dried and powdered the petals and mixed them with yellow cornmeal to use as a face powder for the Basket Dance. Carved wood sunflower designs that were probably used in ritual ceremonies have been found in prehistoric sites in Arizona. Carved and painted sunflowers have also been found on ancient Pueblo pots and kiva walls. The flower heads yield dyes in shades of yellow, orange, and green, while the shells offer a purple or dark red dye.

Among its wildlife consumers, birds are by far the most common, arriving en masse to harvest the maturing seeds. Sunflower is host during its flowering phase to more polite creatures, such as ants and ladybugs, which feed on the aphids on the buds and flower heads. The plant is a nectar source for monarch (*Danaus plexippus*) and bordered patch (*Chlosyne lacinia*) butterflies, and it is host to larvae of the latter, which leave the plants with seriously chewed leaves in the fall.

ASTERACEAE (Sunflower Family)

Hymenopappus flavescens

YELLOW WOOLLY-WHITE, YELLOW PLAINSMAN

The name "yellow woolly-white" sends a confusing color message. Woolly-white gets its common name from the fine white hairs that clothe its stems and leaves, while the flowers are "yellow" (*flavescens*). The seeds are covered with bristly hairs and topped with translucent

Hymenopappus flavescens, yellow woolly-white; mature seed heads with hairy seeds and scaly pappi.

scales (see photo), as indicated by the genus name, *Hymenopappus* (membrane bristle).

Yellow woolly-white is a drought-tolerant biennial that grows in disturbed sandy or limestone soil in the deserts and foothills from 3,500 to 5,000 feet elevation. Woolly-white reaches three to four feet tall, and with the finely divided leaves along the stem and the twinkling yellow discoid flower heads, it adds a beguiling, lacy appearance to the scenery, especially where there is a large grouping. Yellow woolly-white blooms from May to September and adds texture and height to a natural or waterwise landscape.

ASTERACEAE (Sunflower Family)

Hymenoxys odorata

PUNGENT BITTERWEED, WESTERN BITTERWEED, BITTER RUBBERWEED

The Greek word *hymenoxys* (sharp membrane) references the sharp-tipped pappus scales on the achenes. The finely divided leaves and bright yellow flower heads have small glands containing "pungently aromatic" sesqui-terpene compounds, recognized by the species name, *odorata*.

This poisonous annual weed is associated with disturbed land and overgrazed grasslands from 3,500 to 6,000 feet elevation and is the cause of great economic loss to the sheep industry in the Southwest, especially in Texas.

The two main sesquiterpene lactones, hymenoxon and hymenovin, are toxic to livestock, especially to sheep. Hymenoxon, the most important toxin, is found in greater concentrations in the juvenile and flowering stages than in the mature adult. Although sheep generally find it distasteful, they will eat it in the winter and early spring when little else is available. This coincides with the period of the highest concentrations of toxins, which inflame the stomach, damage the kidneys and liver, and reduce the ability of the blood to clot. The poisoning can be acute as well as chronic, since the effect is cumulative, and the outcome is generally fatal. Oddly enough, sheep seem to develop a fondness for the pungent flavor, not unlike the human predilection for wasabi or fiery hot sauce!

Both hymenoxon and hymenovin show antimicrobial activity and have a mutagenic effect on bacteria.

ASTERACEAE (Sunflower Family)

Isocoma pluriflora

JIMMYWEED, RAYLESS GOLDENROD, SOUTHERN GOLDENBUSH

Jimmyweed's discoid flowers are arranged in flat, golden, goblet-shaped flower heads clustered at the ends of twenty-five to fifty vertical stems of roughly the same length. This description is succinctly summarized in its scientific name: *Isocoma* (equal tuft of hair) *pluriflora* (many flowers). Jimmyweed is toxic to livestock, causing them to "get the jimmies," or to tremble violently.

Isocoma pluriflora, jimmyweed; whole plant view at White Sands National Monument.

Isocoma rusbyi, Rusby's goldenbush; note flower clusters with more florets than *I. pluriflora*.

Jimmyweed is a small perennial shrub with resinous leaves that are dotted with glands. It flowers from April through October, depending on the latitude, and grows in a variety of soil types, often in sand or gypsum, up to 5,000 feet elevation. It is commonly associated with *Larrea* (creosote bush) or with *Ericameria* (rabbitbrush), which it superficially resembles. Its attractive color and shape qualify it for a natural landscape setting.

The leaves are usually sticky with resin, which contains tremetol, a mixture of benzofuran ketones that is toxic to sheep, goats, cattle, horses, pigs, and humans. Consumption of as little plant material as 1 percent of body weight over a one- to three-week period is toxic. The poison causes degeneration and necrosis of the skeletal muscles, exhibited by inactivity followed by muscle tremors when the animal is forced to move. Asymptomatic nursing mothers can pass the toxin through their milk, affecting their young, as well as humans who may drink the milk.

The Navajo made a lotion from jimmyweed to heal a baby's navel. The Pima chewed the leaves to calm coughs and made a poultice to relieve muscle pain.

Isocoma rusbyi (see photo) is a related species found in the Four Corners area that resembles *I. pluriflora* but has about twice as many florets in the flower head.

ASTERACEAE (Sunflower Family)

Packera multilobata (syn. *Senecio multilobatus*)

LOBELEAF GROUNDSEL, BASIN BUTTERWEED

John Packer, a contemporary Canadian botanist, has been honored with this genus name. The edges of the leaves are "many lobed," as described by the species name, *multilobata*. The genus resembles *Senecio*, from which it was separated in 1981. However, not all botanists concur with this separation, and many still recognize it as *Senecio*.

This taprooted species is prevalent in northwest New Mexico, growing as a biennial or short-lived perennial in dry rocky or sandy soil in sagebrush to subalpine communities. The bright yellow flowers that appear in spring are replaced by the "gray hair" of mature, fluffy-topped achenes by early summer.

The Ramah Navajo make a tea from lobeleaf groundsel to relieve menstrual pain. In the past, the Yavapai of central Arizona made a tea from the leaves and stems and inhaled the vapor as a cold remedy. They used the leaves to treat skin sores and the roots to relieve stomachache and as a wash for sores resulting from gonorrhea.

Lobeleaf groundsel is being investigated for use in seed mixtures for land reclamation projects. Livestock and wildlife are known to graze it. However, it contains pyrrolizidine alkaloids that, while protecting it against herbivorous insects, cause liver damage in cattle and horses.

ASTERACEAE (Sunflower Family)

Psilostrophe tagetina

WOOLLY PAPERFLOWER

When Augustin Pyramus de Candolle was describing this species in the early 1800s, he used the term *Psilostrophe* for the "naked," or smooth, bract-free, surface of the receptacle (the base to which the flowers are attached). The species name indicates that it was named for a similarity to *Tagetes*, or marigold. Both etymologies are vague in comparison to the

Psilostrophe tagetina, woolly paperflower; hemispherical profile of whole plant.

straightforward description provided by the common name, which tells us that this plant has woolly hairs covering the leaves and stems and that the blooms have a papery texture.

The perimeter of paperflower's lovely mounded form is covered in lemon-yellow flowers. When they reach their expiration date, they simply become dry, or papery, without losing their shape or much of their visual appeal. To create a perfect replica of the living plant, simply cut a handsome specimen and hang it upside down to dry in a warm place.

Paperflower grows enthusiastically and cheerfully in the most unfavorable conditions, in disturbed limestone, saline, or alkaline soil in desert scrub or grassland communities up to 8,000 feet elevation. The plants are usually biennial or perennial, flowering from spring to the first heavy frost. It reseeds readily.

Paperflower contains sesquiterpene lactones, making it toxic to animals, especially sheep. Young plants are more toxic than older ones.

The toxins in the right dosage can serve as medicines, and the Navajo made a strong tea to relieve stomachache and to act as a purgative. The tea could be gargled for sore throat. For ceremonies, the Navajo used a cold tea for an eyewash. The Zuñi treated rattlesnake bite with a poultice made from the roots.

The Pueblo tribes and the Apache extracted a yellow dye from the flowers.

ASTERACEAE (Sunflower Family)

Ratibida columnifera

MEXICAN HAT, UPRIGHT PRAIRIE CONEFLOWER

The genus name, *Ratibida,* was given in 1817 by C. S. Rafinesque, but the significance of the name has been lost to posterity. The species name, *columnifera,* accurately describes the tall, "columnar" receptacle to which

Ratibida columnifera, Mexican hat; note finely divided leaves.

Ratibida tagetes, green Mexican hat; small plant with tiny ray flowers.

the flowers are attached, mirroring the high crown and broad brim of a sombrero, or Mexican hat.

This drought- and heat-hardy perennial grows in both adverse and favorable soil conditions, in disturbed land along roadsides and in native grasslands up to about 6,000 feet elevation. The leaves are pinnatifid, with deep, narrow lobes, as shown in the photo; this characteristic helps distinguish it from similar-appearing cousins. Flowers are abundant from early spring to fall. Mexican hat has become popular in natural plantings for its cheerfully dancing petals and maintenance-free care. Its sole detraction is that it reseeds in great abundance, and some thinning may be necessary to prevent a garden takeover.

The Acoma and Laguna rubbed crushed leaves on the breasts of nursing mothers to help them wean their babies. Also in the nursery department, the Lakota used the flower head as a nipple. The Dakota used the flower heads to treat wounds and relieve chest pains, and they made a beverage from the leaves. The Cheyenne treated snakebite and a poison ivy rash with a strong brew made from the leaves and stems. The Zuñi used Mexican hat as a ceremonial emetic for internal cleansing.

Yellow and green dyes can be extracted from various parts of Mexican hat.

Mexican hat is generally unpalatable to livestock, but deer sometimes use it for browse, and birds and small mammals use it as a minor food source. It establishes easily from seed, making it useful for land reclamation or prairie restoration.

Mexican hat has a little cousin, *Ratibida tagetes*, which has a definite family resemblance but on a much-reduced scale, as shown in the photo. The Ramah Navajo hold this species as one of the Life Medicines. In the past, they made a tea from the leaves to treat stomachache, coughs, and fever. A tea made from the roots helped with birthing injuries.

ASTERACEAE (Sunflower Family)

Rudbeckia laciniata

CUT-LEAF CONEFLOWER, GREEN-HEAD CONEFLOWER, GOLDEN GLOW

Linnaeus named this genus after his mentors and financial supporters, Olof Rudbeck, Sr. and Jr., who were professors of medicine in Sweden. Olof Sr. turned to botany in his older age and eventually established a botanical garden in Sweden, which, not surprisingly, he named the Linnaean Garden. Cut-leaf coneflower's species name reflects the common name: the leaves are deeply divided, or "cut" (*laciniata*).

Rudbeckia laciniata, cut-leaf coneflower; droopy ray flowers indicate flower heads past their prime.

This eight-foot-tall perennial inhabits riparian communities along streams from 5,000 to 8,500 feet elevation. It propagates somewhat aggressively by long, slender rhizomes. Five-inch-wide flowers bloom from July through September. The long, yellow ray petals tend to droop like droll dog ears when the flower is young, then flop entirely as the flower grows old.

The Cherokee ate the leaves as spring greens, but note that the plant may be toxic if consumed in large quantities. The Navajo used it to treat indigestion and colds. Today, as in the past, the Navajo use it in prayers for rain. Its Navajo name indicates that it was a "big arrow poison antidote."

Cut-leaf coneflower is a nectar host to many butterflies and moths, including the white-spotted black diurnal moth (*Gnophaela latipennis*), shown in the main photo. In a moisture-favored landscape, this is a striking plant in a group but needs enough area to accommodate its tendency to spread.

ASTERACEAE (Sunflower Family)

Senecio flaccidus var. *flaccidus* (syn. *Senecio douglasii*)

THREADLEAF GROUNDSEL, THREADLEAF RAGWORT

The image conjured up by the genus and species names is that of a "droopy" (*flaccidus*) "old man" (*senex*), in reference to the white hairs attached to the tops of the achenes. The leaves are usually divided into three very thin sections, as the common name "threadleaf" indicates. A

Senecio flaccidus var. *flaccidus*, threadleaf groundsel; various stages of maturity.

Senecio eremophilus, desert ragwort.

distinguishing characteristic of this species is the dense covering of the leaves and stems with matted white hairs.

The plant is a short, shrubby perennial that grows in rocky or sandy soil along roadsides and in overgrazed rangeland from 4,000 to 7,000 feet elevation. Blooms are open for business from May through October, and the flowers quickly mature into fluffy achenes, giving the plant a frumpy appearance.

The Navajo made a medicine from the leaves for arthritis and rheumatism that was taken orally or used as a poultice. To treat skin sores, the Navajo boiled the plant and allowed the steam to encounter the sores, while the Hopi used pounded leaves for the same purpose. The Jemez steamed the plant over hot coals to stimulate a person who felt weak.

The Navajo used the tea before sweat lodge ceremonies to "get a good voice for the Night Chant." They rubbed the plant over prickly pear fruits to get rid of the spines. The Acoma and Laguna put threadleaf groundsel in their shoes to keep their feet from sweating. They also used it to keep bedbugs at bay and mixed it with *Artemisia* and deer marrow to make a salve.

The leaves are toxic to cattle and horses, with sheep and goats less affected. The plant contains pyrrolizidine alkaloids, including jacobine and seneciphylline as the primary toxins; these toxins increase in concentration under drought stress. Acute poisoning is rare; chronic effects are more common when cattle and horses consume small amounts over a period of time. It may take a real Sherlock Holmes to figure out the source of an animal's ill health because six months may elapse before symptoms appear. The alkaloids cause severe and permanent liver damage, for which no treatment is available.

Senecio eremophilus (see photo) can be distinguished from its more than one thousand cousins by its large, deeply incised leaves and dark, mascara-tipped phyllaries. It grows in rocky soil in mountain and subalpine communities and along roadsides throughout the western half of New Mexico.

ASTERACEAE (Sunflower Family)

Solidago velutina

GOLDENROD, THREENERVE GOLDENROD

Goldenrod's reputation as a healing plant motivated Linnaeus to give it the genus name *Solidago*, meaning "to make whole." The species name, *velutina*, means "velvety," in reference to the fine hairs on the leaves and stems. The leaves have three prominent veins on the underside, as one of its common names indicates. The name "goldenrod" gives an accurate visual image of the yellow flowers arranged in loose spikes on vertical stems.

Goldenrod is a perennial, propagating mainly from rhizomes and spreading aggressively. It grows in waste areas, along roadsides, and in open fields and clearings in forests. While it can tolerate dry soils, it is typically found in moist conditions. The flowers appear from July through the fall.

This and other *Solidago* species share similar chemistry and therefore have broadly comparable healing properties. *Solidago* contains tannins that have astringent properties for contracting the tissue of superficial skin wounds. The roots have more medicinal value than the stems and flowers. A tea made from the roots was used to treat colds, alleviate toothache, and soothe burns. The roots were also used in preparing a poultice for boils. Bee and ant stings were relieved with a preparation from the flowers, and a tea made from the stems and leaves was used to reduce fever and chest pains. Hopi women used goldenrod to reduce the output of breast milk. The Navajo made a cold tea from this species and included it in a lotion to use against witchcraft.

The Omaha Indians used *Solidago* as a seasonal clock, indicating the time when their corn was beginning to mature and they should return home from their summer buffalo hunt.

Native Americans use the leaves and flowers to make yellow and gold dyes for cotton, wool, and baskets.

Solidago is slightly toxic to livestock, which may eat it in the spring when the plants are immature. Other wildlife, even small mammals, make little use of it for either food or cover. Insects' interest in *Solidago* is another matter altogether, as the flowers are invariably alive with all manner of insects. *Solidago velutina* in particular is a host to the larvae of northern checkerspot (*Chlosyne palla*) and little yellow (*Pyrisitia lisa*) butterflies and is a nectar source for the great purple hairstreak (*Atlides halesus*).

Solidago's handsome showy flowers and ability to attract butterflies and birds make it popular as an ornamental. Be aware that you will soon find it where you did not intend it to be, so plan to circumscribe its rhizomes to control its colonizing propensity.

ASTERACEAE (Sunflower Family)

Sonchus oleraceus

ANNUAL SOWTHISTLE, COMMON SOWTHISTLE

When Linnaeus named this plant in 1753, two notable features seemed uppermost in his mind: the "hollow" (*sonchus*) drinking-straw-like stem, and the plant's popularity as a kitchen vegetable, or "potherb" (*oleraceus*). It is an equally popular menu selection for livestock, and the name "sowthistle" indicates porcine enthusiasm for its culinary delights.

Sonchus oleraceus, annual sowthistle; pinnatifid leaves.

Sowthistle was introduced from Europe by early American settlers and has now spread to all states. Several Southwest Indian tribes use it for food, boiling the tender young leaves (see photo) in salted water with chili to eat as greens. Some rub the leaves and stems between their hands and eat it raw.

Europeans from early Roman times knew sowthistle as a nutritious vegetable, and a diet rich in sowthistle was recommended to men by Renaissance-era herbalists to maintain their "manhood." It has a broad range of curative properties arising from its chemical-rich milky sap. In the 1600s, it was used as a beauty treatment for women's skin, and in modern Europe it is taken as a treatment for cardiovascular problems, including high blood pressure, and to bring down fever. Recent research has shown that it does have anti-inflammatory and fever-reducing properties.

Among the Southwest Indians, a tea was used to reactivate the menstrual cycle, effectively serving as an abortion agent. The Pima used the gum from the sap to help cure opium addiction, as was also done in China.

Not everyone is so enamored with sowthistle, as it is considered to be weedy and invasive in many states. It also plays host to a variety of garden pests, particularly aphids, which thrive on its juices and leave their own sticky secretions, which catch dust and fuzz.

ASTERACEAE (Sunflower Family)

Taraxacum officinale

DANDELION, BLOWBALL

Dandelion is a native of Eurasia but has been naturalized in many countries. The humble, hated weed is a veritable Cinderella in the world of plants, complete with golden hair! Despite dandelion's long, auspicious history trailing clouds of glory in the food and medicine departments, the current generation seems churlishly unappreciative of its great value.

The genus name, *Taraxacum*, originated in the Persian *talkh chakok*, meaning "bitter herb." Europeans valued *Taraxacum officinale* so much that it was the "official" (*officinale*) word in medicine. "Dandelion" is a corruption of the French phrase *dent-de-lion*, meaning "lion's tooth," in reference to the deeply divided, "dentate" leaves.

The flowers quickly mature into seeds topped with white fluffy hairs. Children around the world, both young and old, make a game of giving a big puff to these fluffy heads, or "blowballs," to see them float into the air. The rapid metamorphosis from flower to fruit gave rise to a legend in which the dandelion represented a golden-haired maiden who in a short period of time turned into an old woman with white hair—the very parable of fleeting youth.

In many regions, dandelions have three sets of chromosomes, instead of the typical two, and produce a significant proportion of their seeds by apo-

Taraxacum officinale, dandelion; lawn view.

mixis, without fertilization of the ovary by pollen. The apomictic populations seem to be better colonizers than the sexually reproducing ones.

Young dandelion leaves are high in vitamins A and C and make a tasty cooked or raw green vegetable. The flowers are delicious battered and fried (but then, what isn't?) and served in much the same way as fried okra. For a sweeter treat, roll the fried flowers in powdered sugar and serve them with syrup. The flowers may also be fermented to make dandelion wine. The Apache used the flowers to flavor their drinks and believed that the blooms made them stronger. Eating the roots is a good natural laxative due to the presence of inulin, which serves as a source of fiber to keep things moving along. Some pioneers are said to have roasted the roots and used them as a coffee substitute.

The leaves and roots have a mild diuretic effect. The Tewa and Navajo found that swelling and bruises could be effectively treated with a poultice of cooked dandelion leaves. Ramah Navajo women drank dandelion tea during a difficult labor to speed the delivery of the baby. Dandelion sap is said to be an effective wart remover, as well as a treatment for ringworm.

Livestock and wild animals also rely on dandelion for food, finding it more palatable in the pre-bloom stages, as the leaves become somewhat bitter as they age.

ASTERACEAE (Sunflower Family)

Thelesperma megapotamicum

COTA, RIO GRANDE GREENTHREAD, INDIAN TEA

Cota is one of the most important plants for Southwest Native Americans, valued for its beverage and medicinal properties. *Thelesperma* (nipple seed) refers to the nipple-shaped bumps on the surface of the achenes. The species name, *megapotamicum*, means "big river," referring to the Rio Grande.

Cota is a slender, knee-high, usually very branching annual with thumb-size flower heads atop skinny, leafless stalks. The leaves are mostly basal. It has no ray flowers, or "petals," and blooms from April through October. It grows on disturbed land in a variety of soils and is found throughout New Mexico in desert scrub to pinyon-juniper communities from 4,000 to 7,000 feet elevation.

Cota is widely sought to brew a popular tea enjoyed by Native Americans throughout the Southwest. Many keep dried cota in the pantry the same way other people keep Lipton tea bags. The stems are harvested by breaking them off aboveground so the plant can regrow, then they are folded into three-inch-long bundles and tied for drying.

Thelesperma megapotamicum, Indian tea; seed head showing two-pronged pappus atop each seed.

Not only is cota tea a refreshing beverage, but for the Navajo it also has medicinal value as a diuretic and to relieve stomach disorders, muscle cramps, and toothache. The Cochiti used a strong cota tea to reduce fever. Soaking irritated skin in cota tea is soothing.

Cota yields rich, permanent orange, gold, and reddish-brown dyes that the Hopi, Navajo, and Jicarilla Apache use to color cotton, wool, and baskets. The Navajo make a red-brown dye from the flowers to use for baskets and fabrics, and Navajo wool weavers create various shades of orange by mixing cota with dock (*Rumex*, p. 154) and mountain mahogany bark (*Cercocarpus*, p. 332). The roots yield an orange-yellow color for dying wool.

ASTERACEAE (Sunflower Family)

Thymophylla acerosa (syn. *Dyssodia acerosa*)

DOGWEED, PRICKLYLEAF DOGWEED

Dogweed's little leaves, resembling those of thyme, inspired its genus name, *Thymophylla* (thyme leaf), and their sharp, "needle-like" tips are highlighted by the species name, *acerosa*. The leaves have small, usually translucent, glands brimming with an aromatic resin with a terpene scent, which inspired a previous genus name, *Dyssodia*, meaning "stench." The name "dogweed" refers to its original habitat in the disturbed soil of prairie dog towns.

Dogweed is a perennial, forming small shrubs, and is most at home

Thymophylla acerosa, dogweed; note needle-tipped leaves.

in dry sandy, rocky caliche, or gypsum soil. It grows throughout southern New Mexico in desert scrub and pinyon-juniper communities from 3,500 to 6,000 feet elevation. The inch-wide yellow flowers of *Thymophylla acerosa* and the slightly smaller heads of its cousin, *T. pentachaeta*, as shown in the photos, bloom from summer through fall.

The Acoma and Laguna found that they liked the pungent leaves mixed with tobacco—an early version of wintergreen Skoal! The Isleta made a decoction of dogweed leaves to bathe a person suffering from fever. When the people of the Omaha and Sioux tribes had a bad headache, they crushed the dried plant into a fine powder and inhaled it to induce nosebleeds in order to relieve the pain and pressure.

Thymophylla pentachaeta (common name "parralena") has red glands on its leaves and phyllaries—lined up like tiny red chicken pox pustules in orderly ranks—that bear a pungent resin with an "eau de marigold" fragrance. The flowers are exserted well above the leaves on slender flower stalks and are distributed all around the plant, giving it a "professional florist arrangement" look. The Kayenta Navajo took advantage of

Thymophylla pentachaeta, parralena.

parralena's strong odor to make a lotion with which they covered themselves before a hunt to avoid detection by their prey.

Both species host larvae of the dainty sulphur butterfly (*Nathalis iole*). Birds and small mammals feed on the seeds. Dogweed is recommended for landscaping because of its extreme drought and heat tolerance. Both species make good border plants and ground cover.

ASTERACEAE (Sunflower Family)

Tragopogon dubius

GOAT'S BEARD, JOSEPH'S FLOWER

This plant has a handsome yellow flower, but it's the seed head with the white Afro that gets all the attention. When Linnaeus named the plant in the 1700s, he called it *Tragopogon*, meaning "goat's beard." The species name means "doubtful," but its significance is obscure. The common name "Joseph's flower" refers to the tradition that Joseph, husband of Mary the mother of Jesus, had a long white beard.

Goat's beard blooms in the spring and early summer, throwing open its yellow ray flowers and very long phyllaries early in the morning. A late-rising botanist will be greeted with closed flowers, since they keep limited business hours. The stems and leaves exude a milky sap when broken. The plants grow in disturbed soil along roadsides or abandoned habitations in various plant communities up to about 8,000 feet elevation.

Tragopogon dubius, goat's beard; seeds with feathery pappi ready for takeoff.

Goat's beard is an immigrant from Europe that has settled throughout North America. Europeans valued its medicinal properties, especially for soothing digestive difficulties such as heartburn and for treating liver problems and pleurisy. Early North American settlers and Native Americans also welcomed it into their medicine chests as it spread across the country.

Goat's beard's leaves, stems, and roots are all edible when young. They may be eaten raw in a salad or boiled as a cooked vegetable. Early pioneers sometimes roasted the roots and used them as a caffeine-free, "buzzless" coffee substitute. Native Americans collected the sap and allowed it to coagulate to use as chewing gum.

American Indians found that eating goat's beard roots kept them "regular," thanks to the presence of inulin, which not only increases laxation but also aids in regulating blood sugar in diabetics. The Ramah Navajo still use it as a ceremonial emetic to clean out the body prior to entering the sweat lodge. The Navajo give large amounts of the tea to horses with internal injuries.

Large wildlife, including bighorn sheep, deer, elk, pronghorn, and wild horses, consume goat's beard, as do small mammals such as gophers and mice; birds, including various grouse species; and domestic cattle and sheep.

ASTERACEAE (Sunflower Family)

Uropappus lindleyi

SILVERPUFFS

Silverpuffs' flowers are not anything that would attract an audience, but the seed heads are guaranteed to draw attention. The achenes are longer than typical for the sunflower family and are outfitted with five long, wide pappus "tails" (*uro*), giving it a starburst appearance with an overall startling effect. The species name honors John Lindley, a nineteenth-century botanist, indefatigable writer of botanical treatises, and proponent of a classification system based on family groups.

Uropappus is a monotypic genus, meaning it has only one species. This annual grows in arid southwest New Mexico in desert grassland and chaparral communities up to 6,000 feet elevation. The flowers close early in the day, and the stems contain a milky latex.

Uropappus lindleyi, silverpuffs; seed head amid buds.

Uropappus lindleyi, silverpuffs; seeds with large, star-shaped pappi.

ASTERACEAE (Sunflower Family)

Verbesina encelioides

COWPEN DAISY, GOLDEN CROWNBEARD

Cowpen daisy's scientific name promotes its resemblance to other species rather than its own outstanding features. The genus name indicates that its leaves resemble those of *Verbena*, and the species name tells us that the plant has the look of *Encelia*, or brittlebush, with which it is easily confused. Its barnyard name, "cowpen daisy," comes from its habit of growing in disturbed areas such as livestock enclosures, railroad rights-of-way, and roadsides, as well as from the "eau de bovine" scent released when the leaves are rubbed.

The leaves and stems are hairy, and the young leaves look especially well groomed, with all the hairs lying down in the same direction. This three-foot-tall annual grows throughout New Mexico in desert scrub, grassland, and pinyon-juniper communities below 7,000 feet elevation. Blooms cover the plant from early spring through fall, and seed production is prolific. Cowpen daisy is classified as weedy or invasive in many states because of the great fecundity of its seeds.

Verbesina encelioides, cowpen daisy; characteristic leaves.

Cowpen daisy's alkaloid compounds have anti-inflammatory and antibiotic properties that render it useful for treating skin sores, wounds, and bites. The Kayenta Navajo make an infusion to treat spider bites, both drinking the tea and using it to wash the wound. The Ramah Navajo and Hopi make a poultice of the leaves to cover rattlesnake bites. A tea made with the leaves helps relieve gas and bloating, and the leaves also have been used to poultice hemorrhoids and skin ulcers.

Small mammals and terrestrial birds use cowpen daisy as a food source, and a number of butterflies enjoy its nectar. Larvae of the bordered patch butterfly (*Chlosyne lacinia*) use it as food.

Cowpen daisy's ability to bloom continuously throughout the hot, dry summer and its attractiveness to butterflies make it a desirable member of a waterwise landscape. It also has good deer resistance. Be aware that its reseeding ability and robust growth habit will overwhelm less competitive neighbors.

ASTERACEAE (Sunflower Family)

Xanthisma spinulosum (syn. *Machaeranthera pinnatifida*)

YELLOW SPINY DAISY, LACY TANSYASTER, SPINY GOLDENWEED

Xanthisma means "that which is dyed yellow," referencing the lemon-yellow flowers covering this plant. *Spinulosum* means "with many little spines," describing both the margins of the leaf lobes and the phyllaries, which are tipped with small white bristles that are prickly indeed when the plant is handled. The common name "yellow spiny daisy" combines both of those concepts.

Yellow spiny daisy is a remarkably variable species in every aspect of its growth habit across its broad geographical range. This has led to equally remarkable confusion among taxonomists about whether the differences justify separation of the plants as species, varieties, or races.

The plants, meanwhile, continue growing cheerily in dry sandy soil in open areas of desert grasslands and pinyon-juniper communities up to about 7,000 feet elevation. Each little branch of the plant ends in a yellow flower. Bloom begins in the spring and continues through the fall to frost, with a concurrent continuous cycle of flowers and seed heads. After the first beautiful flush of flowers in the spring, the plant begins to look more and more unkempt. It exudes a sticky resin that collects dirt and bits of fuzz, adding to its down-at-the-heels look.

The Navajo use yellow spiny daisy as a treatment for headache. Other Native Americans have made poultices of the crushed leaves to heal wounds and chewed the roots to reduce toothache.

Yellow spiny daisy is very suitable for a native plant landscape, but it reseeds prolifically, and its numerous progeny may outcompete other desirable plants in the next season.

ASTERACEAE (Sunflower Family)

Zinnia grandiflora

ROCKY MOUNTAIN ZINNIA, PLAINS ZINNIA

Zinnia was named by Linnaeus to honor Johann Zinn, an energetic German who, in his short thirty-two years of life in the mid-1700s, was an ophthalmology professor and director of the botanical gardens in Göttingen. Not only did he provide an accurate description of the eyeball's nerves and blood vessels, but he also traveled to Mexico, where he collected seeds of what is now the garden plant *Zinnia*.

The species name, *grandiflora*, meaning "grand flower," is aptly applied to this plant, whose stunning color and pleasing shape impart great land-

Zinnia grandiflora, Rocky Mountain zinnia; whole plant.

scaping value both in the wild and in the garden. This perennial is found throughout New Mexico in dry calcareous soil in grassland, chaparral, and pinyon-juniper communities from 4,000 to 6,500 feet elevation. The plants flower from May through October, with the petals becoming papery and dry in maturity.

Native Americans used zinnia to control diarrhea, and they crushed the entire plant and mixed it with salt to treat rheumatism. The Ramah Navajo made a tea to relieve heartburn, clear a stuffy nose, and soothe a sore throat. The Acoma and Laguna still prepare hot zinnia tea for anyone suffering from kidney ailments, and they bathe in the tea to prevent excessive sweating. The Zuñi inhale zinnia smoke in a sweat lodge to bring down fever. They wash their irritated eyes with a cold tea made from the blossoms. In the past, ambitious parents fed the plant to their children to encourage them to begin to speak at a younger age.

The plants are extremely drought tolerant and bloom even in stressful conditions. Natural landscapers in the Southwest use zinnia as a border or ground cover. Propagation is primarily by rhizomes, which enlarge the colonies, making Rocky Mountain zinnia useful for erosion control and restoration of damaged land. It has perfect manners and rarely crowds out neighboring plants in a landscape setting.

Insects of all sorts, including butterflies and moths, feed on the pollen and nectar. Two shown here are the caterpillar wasp (*Ammophila* spp.) and, in the main photo, checkered white butterfly (*Pontia protodice*). Small

Zinnia grandiflora, Rocky Mountain zinnia; with caterpillar wasp (*Ammophila* sp.).

birds, including doves, finches, and quail, consume the seeds. Rocky Mountain zinnia has good deer resistance.

Also shown here is its cousin *Zinnia acerosa,* which shares many of the same characteristics but has white flowers and a limited range in southern New Mexico.

Zinnia acerosa, desert zinnia.

BERBERIDACEAE (Barberry Family)

Berberis fremontii (syn. *Mahonia fremontii*)

FREMONT'S BARBERRY, HOLLY GRAPE, AGRACEJO

The genus name, *Berberis*, is derived from an Arabic word for the plant's colorful, tart fruits. John Charles Frémont, who lived through most of the 1800s and made a number of plant-collecting and cartography expeditions to the West Coast, is honored with the species name.

Berberis fremontii is found in northwest, central, and southern New Mexico from 5,000 to 7,000 feet elevation. The ten-foot-tall bushes look unkempt except when they are covered in glorious masses of yellow flowers in the early spring, from February to April. They set small, shiny, juicy red berries from May to August. The Apache ate these berries fresh and cooked in a jelly.

The plant's leaves and roots contain bitter alkaloids, including berberine, which is antimicrobial, anti-inflammatory, and a fever reducer. The roots, colored yellow by berberine, have been used to treat toothache and stomachache. This plant has also traditionally been used to treat chronic liver problems. The Apache soaked wood shavings in water and used the resulting tea as an eyewash. The Navajo treated stomachache, heartburn, and other digestion problems with Fremont's barberry and mixed it with other plants to treat spider bites.

The Apache decorated their skin for certain ceremonies with the purple berry juice. Root shavings produce a yellow dye for wool and hides. Tools and small carvings are made from the yellow wood.

The plant's leaves and yellow flowers are attractive in natural landscaping projects and for use as a barrier hedge. In the fall, the leaves turn

Berberis fendleri, Fendler's barberry; note simple leaves on spiny branches.

Berberis fendleri, Fendler's barberry; red berries.

yellow and red. Plants are easily established from cuttings and from seed and may be useful for rehabilitating damaged sites. Many birds and small mammals consume the berries, and the bushes provide cover for many wildlife species. The flowers attract a variety of insects, including honeybees, and the honey made by bees feeding on Fremont's barberry is highly valued.

The *Berberis* genus is the alternate host to the wheat disease *Puccinia graminis*, or stem rust fungus, which causes major economic losses if not controlled. Movement of *Berberis* into the northern Great Plains and north-central states, where most U.S. wheat is produced, is therefore forbidden.

A relative, *B. fendleri*, also shown here, grows in northern New Mexico at higher elevations than *B. fremontii*, from 6,000 to 8,500 feet, in canyons, on hillsides, and along streams.

Also pictured is another cousin, *B. trifoliolata*, a dominant member of desert and grassland communities in southern New Mexico at less than 3,500 feet elevation.

Berberis trifoliolata, agarita; note trifoliolate leaves and immature berries.

BORAGINACEAE (Forget-me-not Family)

Lithospermum incisum

FRINGED PUCCOON, INDIAN PAINT, NARROWLEAF STONESEED

The scientific names, *Lithospermum* (stone seed) and *incisum* (cut), indicate that the seeds are rock hard and the leaves are deeply cut. The common name "puccoon" is of Native American origin, referring to plants that yield dyes; in this case, bluish-violet colors can be extracted from the root. The plant is covered with small, tidily "combed" hairs lying in the same direction.

Puccoon is usually about a foot tall, with its frilly-fringed yellow flowers appearing from May to July. Early-season flowers put on a show, but later-season flowers are small and pale and may never open. Puccoon grows throughout most of New Mexico in grasslands and pinyon-juniper communities from 4,000 to 7,000 feet elevation.

Native Americans dug up the short, fleshy root to eat as a cooked vegetable. The root was also used as a contraceptive. People wanting to space the arrival of their children drank a cold infusion of puccoon root every morning. The validity of this practice has been corroborated by experiments with rabbits showing that puccoon inhibited the action of gonadotropins in the ovaries. Mice fed puccoon extracts showed a reduction in the size of their reproductive organs, as well as the thymus and pituitary gland.

Puccoon contains tannins that promote the healing of wounds. The Zuñi made a salve of the powdered root to treat abrasions and skin infections. They also used puccoon tea to treat sore throat and kidney problems. The Navajo used puccoon to treat infections of the navel in newborns and made a cold tea from the root and seeds to soothe irritated eyes.

BRASSICACEAE (Mustard Family)

Descurainia pinnata

WESTERN TANSYMUSTARD

Descurainia was named in honor of the eighteenth-century French botanist, pharmacist, and physician François Descourain. The species name, *pinnata*, describes the leaves as "feathery," or "pinnate."

Descurainia pinnata, native to North America, can be distinguished from its cousin *D. sophia*, a weed introduced from Europe, by its half-inch-long, club-shaped seedpods. This annual reaches three feet tall and is usually covered with fine white hairs. It grows throughout New Mexico in disturbed soil in open areas from 4,000 to 7,500 feet elevation. Western tansymustard is an indicator plant for ruin mounds. It is often found growing above buried prehistoric rooms, because weathering of adobe walls or the clay mortar in stone walls increases the clay content of the soil and therefore holds more moisture that encourages more lush plant growth above the buried ruins.

Tansymustard is one of the first plants to spring up at the end of winter, and it provided tasty greens, eaten raw or cooked, to Native Americans. The tiny mustard seeds were a food source since prehistoric times into the modern era. At Salmon Ruins in Bloomfield, New Mexico, archaeologists found tansymustard seeds among the pueblo ruins and in feces. Two gallons of tansymustard seeds were recovered from a ruin near Aztec, New Mexico.

The Ramah Navajo made cakes from the seeds, which were first parched, then ground into flour. Their zingy taste comes from the sulfur-containing glucosinolate compounds, which have anticancer and anti-inflammatory properties and particularly help to protect against colon cancer. The Ramah Navajo also used the crushed, boiled plant as a poultice to relieve toothache. The Kayenta Navajo made a lotion to rub on the neck to relieve sore throat.

The mustard family contains glycosides with antifungal properties. Without knowing this "scientific fact," the Acoma and Laguna soaked seed corn in an infusion of mustard leaves to help control seed-destroying fungi during germination. Then they buried the leaves as fertilizer with the seed. Mustard leaves were also stored with harvested corn as a fungicide to prevent growth of mold.

The yellow flowers of *D. pinnata* were mixed with dark iron pigment to make a black color for pottery decoration.

Tansymustard is a minor food source for large mammals and terrestrial birds, and it serves as a larval host for desert orangetip (*Anthocharis cithara*) and spring white (*Pontia sisymbrii*) butterflies.

BRASSICACEAE (Mustard Family)

Erysimum capitatum

WESTERN WALLFLOWER, SANDDUNE WALLFLOWER, PRAIRIE-ROCKET

The medicinal value of western wallflower is recognized in the genus name, from the Greek word *eryso,* "to ward off" or "to cure." The species name, *capitatum,* refers to the "headlike" cluster of the flowers.

Western wallflower is highly variable in size and flower color throughout its range. It is a biennial or short-lived perennial, found across New Mexico in pinyon-juniper to spruce-fir zones from 7,000 to 11,500 feet elevation. The showy flowers bloom from May to July in colors varying from yellow to orange to maroon.

The Ramah Navajo inhaled the crushed leaves to relieve headache pain, and they used the plants as a ceremonial emetic. In the days before dentists, they made a hot poultice of the roots to soothe toothache. The Zuñi ate the flowers and seedpods to cure stomachache. The Acoma and Laguna used wallflower leaf poultices to reduce swelling.

Native Americans dried the flowers and ground them into a powder to make a yellow paint used in ceremonies.

The handsome yellow flowers and the long, stiffly upright fruits are eye-catching in a natural landscape. Wallflower seed is commercially available.

Pikas and rockchucks feed on wallflower's stems and leaves, and elk and bighorn sheep consume it to some extent at higher elevations. The caterpillars of cabbage white (*Pieris rapae*) and orangetip (*Anthocharis* spp.) butterflies feed specifically on plants of the mustard family, while adults nectar on flowers from other plant families as well as the mustard family.

BRASSICACEAE (Mustard Family)

Physaria fendleri (syn. *Lesquerella fendleri*)

FENDLER BLADDERPOD, POPWEED

The genus name is based on the Greek word *physa,* meaning "bladder," in reference to the plant's spherical, bomb-shaped fruits, with a remnant of the style serving as the "fuse." The species name honors Augustus

Physaria fendleri, bladderpod; flowers and seedpods.

Fendler, a nineteenth-century German botanist who collected extensively in New Mexico.

These low-growing perennials may reach one foot tall, with silvery stellate hairs covering the leaves and stems. *Physaria* is one of the first to bloom in the spring, carpeting large areas with bright yellow flowers that transform into round fruits that pop when squeezed, giving rise to another common name, "popweed." It is commonly found throughout New Mexico on rocky, calcareous slopes or in sandy soil from 3,500 to 7,000 feet elevation. With adequate rain, Fendler bladderpod can bloom through most of the year.

The Navajo make a tea from the leaves to wash spider bites. In the past, the Acoma and Laguna crushed the plant, mixed it with salt, and used it to reduce swelling.

The seeds of all *Physaria* species contain an oil that has a unique fatty acid composition of high commercial interest. Fendler bladderpod accumulates the highest amount of lesquerolic acid, up to 30 percent of its oil, which is targeted as a replacement in several applications for the extremely valuable castor oil, the production and processing of which is restricted in the United States because of the deadly toxin ricin in the castor plant's seeds. *Physaria* oil may prove to be economically important in the manufacture of plastics, coatings, lubricants, and cosmetics. In addition, the seed coat contains a gum, with properties similar to those of xanthan, which can be used as a thickening agent for food and for the recovery of crude oil.

Fendler bladderpod's cheerful flowers and low water and maintenance requirements make it attractive in many types of native landscape

Physaria acutifolia, double bladderpod; seedpods.

settings. Seed is commercially available. Bladderpod's tolerance of alkaline soils has led to its use in the phytoremediation of soil contaminated with selenium, which the plant accumulates in its leaves and stems.

Also pictured here is *Physaria acutifolia*, which has distinctive blue-green "pointed leaves" (*acutifolia*) that form rosettes at ground level. Its seedpods are in pairs, with the remnant of the style in the center. It grows only in the Four Corners area of New Mexico.

BRASSICACEAE (Mustard Family)

Stanleya pinnata

PRINCE'S-PLUME, DESERT PRINCE'S-PLUME, GOLDEN PRINCE'S-PLUME

Thomas Nuttall named this genus for Lord Edward Stanley, 13th Earl of Derby, who was a patron to many naturalists, including John James Audubon. *Pinnata* means "feathered," referring to the deeply pinnately lobed basal leaves.

Stanleya pinnata, prince's-plume; field view.

A sighting of prince's-plume is guaranteed to make you slam on your brakes; it's simply not possible to drive by this plant without stopping for a long, close look at Its Royal Highness. Fortunately, it grows in desert foothill, chaparral, and pinyon-juniper zones, where a quick stop will not create a traffic hazard. It blooms from April through September with bright yellow flowers on two-foot-long racemes. The leaves are blue-green and have a disagreeable smell due to the selenium the plant accumulates, and which makes it toxic.

Despite the plant's dangerous makeup, Native Americans ate it as a cooked green vegetable after boiling it in two or more changes of water to remove the toxic selenium. Mormons trekking to Utah called it "wild cabbage" because of the flavor profile shared by these two members of the mustard family.

The Navajo, who count this plant as one of their Life Medicines, used prince's-plume as a poultice for swollen glands and as an emetic. The Zuñi made poultices of fresh chewed pods to counteract itching. In the years before the advent of modern antibiotics, they dusted the powdered plant over syphilitic sores.

Because prince's-plume is a hyperaccumulator of selenium, which is usually associated with uranium ore, the uranium industry uses the plant as an indicator of the ore's presence.

Such a handsome plant should not be left out of a native landscape or natural garden. Seed is commercially available. Prince's-plume needs dry, well-drained soil and thrives with conservative maintenance.

CACTACEAE (Cactus Family)

Opuntia phaeacantha

PLAINS PRICKLY PEAR CACTUS, TULIP PRICKLY PEAR, BROWN-SPINED PRICKLY PEAR

The first-century Roman naturalist Pliny gave the name *Opountos* to a prickly plant (not a cactus) growing around the Greek city of Opus. Philip Miller adapted the name as *Opuntia* for the cactus group in the 1700s. The species name, *phaeacantha*, is derived from the Greek words meaning "dark spines."

Plains prickly pear is found throughout New Mexico in desert and grassland communities up to 7,500 feet elevation. The flowers bloom from April through June and set fruit during the late summer. Individual plants can live up to twenty years, but through vegetative reproduction their life as a clone becomes perpetual.

Prickly pear pads ("nopales," derived from the Nahuatl [Aztec] word *nopalli*) have long been a major food source throughout the Southwest, both for humans and for animals. Archaeologists have found coprolites (fossilized human feces) that contain prickly pear fibers at several sites, most notably at the Hinds Cave site in the Lower Pecos Canyonlands, where the fiber density in the coprolites indicated that cactus pads were one of the primary food sources at the time. Stacks of desiccated prickly pear pads have been unearthed in caves, possibly having been stored as winter food (see photo). Fresh nopales are still popular and may be purchased in many supermarkets, thankfully with the spines already

Opuntia phaeacantha, plains prickly pear cactus; fruit (tuna).

removed. They may be sliced and cooked by many different methods, including boiling and frying. They are excellent mixed with eggs in a breakfast burrito or omelet. They can also be cubed, rolled in cornmeal, and fried like okra, a fruit with which the cactus shares a slippery similarity. The pads contain mucilaginous (slimy!) complex carbohydrates that lower cholesterol, help regulate blood sugar in diabetics, and provide protection against prostate problems.

The fruits ("tunas," derived from the Taino word) are a source of purple dye and were used to add color to cornmeal porridge. The spines are removed by singeing them in hot ashes or rubbing them off with another plant, such as *Senecio* (see p. 234). The wine-colored fruit juice makes wonderfully appetizing jellies, syrups, candies, and pies.

Prickly pear pads are a practical and portable first-aid kit. The liquid in the pads is soothing for burns and wounds, and for prolonged relief a pad can be split open and laid across the burned area or wound. Juice from the pads has been used to treat urinary tract infections, insomnia, diabetes, arthritis, and headache. Many people keep a jug of cactus juice in the refrigerator to drink as part of their daily health regimen.

The Navajo make rose and pink dyes for wool from the tunas. The Zuñi used the fruits, mixed with *Cleome*, or bee plant (see p. 234), to make a reddish polish for moccasins and for dyeing thread. Consuming copious quantities of ripe tunas will even dye the urine red! The Havasupai used the mucilaginous cactus gel to help prepare pottery clay, and the Ute tempered pots with it before firing. It also served as a glue

Desiccated cactus pads unearthed in a cave in a Lower Pecos Canyonlands archaeological site.

for making clothes and weapons. The gel inside the pads makes a good conditioner for the hair.

Plains prickly pear is an important food source for a wide range of wildlife. Small birds and mammals use it as cover. In a drought, ranchers can burn the spines off to provide prickly pear pads as an emergency forage for livestock. In Mexico, prickly pear is produced commercially as feed for dairy and beef cattle.

Prickly pear's significant role in the ancient and modern history of Mexico is acknowledged by the prominent position of a fruiting plant on the country's national flag.

New Mexico has several species of prickly pear, with different shapes and colors. Shown here are *Opuntia microdasys* var. *rufida* and *O. macrocentra*.

Opuntia microdasys var. *rufida*, blind prickly pear; note fuzzy glochids.

Opuntia macrocentra, purple prickly pear.

CUCURBITACEAE (Gourd Family)

Cucurbita foetidissima

BUFFALO GOURD, COYOTE MELON, MISSOURI GOURD

The first contact with buffalo gourd will be a memorable one: the dreadful smell is not easily forgotten! The species name, *foetidissima,* aptly describes the "extremely fetid" odor emanating from the silvery, triangular leaves. This odor is caused by bitter, herbivore-repellent chemicals in the leaves. The genus name, *Cucurbita,* is Latin for "gourd."

Buffalo gourd is found mainly in the central and eastern half of New Mexico, growing in a variety of plant communities up to 7,000 feet elevation. The vine has a diameter of up to twenty feet, with each runner producing several bright orange-yellow flowers with fused petals. Like many members of the gourd family, buffalo gourd is monoecious, with separate male and female flowers on the same plant. The main photo here shows a female flower and a developing fruit. A male flower is shown in another photo.

Archaeological evidence indicates that people have been using gourds for food for thousands of years. The seeds, which contain 25 to 40 percent oil and 25 to 35 percent protein, were eaten roasted or boiled and sometimes ground into meal. Oil from the seeds is useful for cooking and for hair and skin care. Several groups of Native Americans ate the foul-smelling fruits, which required some processing, since they also contain bitter-tasting and toxic chemicals.

Buffalo gourd's roots are enormous, growing several feet long and weighing up to 150 pounds. Sometimes the root divides so that it resembles a human body. This led people to ascribe healing powers to

Cucurbita foetidissima, buffalo gourd; male flower.

Cucurbita foetidissima, buffalo gourd; mature fruits on vine after frost.

the plant. The root part that corresponded to the injured body part was removed and used as medicine. Taking advantage of the astringent and disinfectant properties of this plant, the Acoma and Laguna boiled the roots to poultice boils and sores, while the Apache used similar poultices for sores on horses. The Tewa pulverized the roots and mixed them with water to drink as a laxative, which could be quite a potent cleanser. Pioneers used an old home remedy for curing sore hooves: they soaked the horse's foot in a bucket of boiled gourds. Some modern-day horsemen use this remedy as well.

The Navajo made an insecticide from the malodorous crushed leaves. Puebloans placed chopped gourds and leaves in all four corners of the house to keep out bugs, and they ground the root to kill bedbugs. An extract from boiled gourd roots was used to kill maggots infesting wounds. Parasitic worms were eliminated by swallowing crushed gourd seeds mixed with water.

In recent times, a commercial insecticide for corn rootworm was based on one of the plant's chemical components, cucurbitacin, which acts as a bait to entice adult beetles to consume the pesticide.

The enormous roots and the outer shell of the fruits contain soapy, foamy saponins that can be used for shampoo, bathing, and laundry. The Papago grated the green fruits into soapy water to bleach clothes. The roots also contain starch, which was used in stiffening fabric.

The dried gourds have a relatively tough outer skin and can be used to make cups, ladles, and other handy items. Native American dancers use them as leg rattles.

FABACEAE (Bean Family)

Hoffmannseggia glauca

HOG POTATO

The genus name honors Count Johann Graf von Hoffmannsegg, a German botanist, entomologist, and ornithologist, whose long life straddled the eighteenth and nineteenth centuries, during which he coauthored a book on the flora of Portugal and established a zoological museum in Berlin. Hog potato leaves have a "gray-green" look, signified by its species epithet, *glauca*.

Hog potato is just plain cute, from its pixy-faced, gland-dotted flowers, its dainty leaves, and its red-spotted seedpods all the way to its little underground "potatoes." The plant, which rarely exceeds a foot in height, has bipinnate leaves and orange-yellow flowering racemes thrust above

Hoffmannseggia glauca, hog potato; whole plant.

the leaves from March through September. It has a sticky feel, thanks to an abundance of little red glands covering the stems and flowers.

Hoffmannseggia glauca, hog potato; gland-studded seedpod.

Part of this plant's charm is its perseverance. Hog potato thrives in adversity—no lap of luxury for this hardy soul, only dry, alkaline desert conditions or, in an urban area, the corner of an asphalt parking lot to push up through, lifting its flowers in triumph.

The thumb-size tuberous swellings on the roots are nutritious and provide food for many animals, most notably hogs, hence the common name. Native Americans dug up these carbohydrate-rich "potatoes" to roast or to store for winter use.

FABACEAE (Bean Family)

Melilotus officinalis

YELLOW SWEET CLOVER, HONEYCLOVER, YELLOW MELILOT

Sweet clover was introduced from Europe in the mid-1600s and spread by beekeepers and farmers who appreciated the high-quality honey

Melilotus officinalis subsp. *alba,* white sweet clover; with Reakirt's blue butterfly.

and forage it provided. In fact, the genus name, *Melilotus*, means "honey legume." The species name, *officinalis,* reflects its historical use as an "official" medicine. A cousin, *Melilotus officinalis* subsp. *alba* (shown here) has white flowers and blooms about three weeks later than yellow clover.

This biennial has deep roots that host *Rhizobium* bacteria, which fix atmospheric nitrogen to add to the soil. Its petite flowers bloom from April through October, with seed set taking place in the second year. It is found throughout New Mexico in all soil types and plant communities up to about 8,500 feet elevation.

Melilotus officinalis produces coumarin, which when converted to dicoumarol is used medicinally as an anticoagulant and as a popular rat and mice killer known by the name warfarin. Improperly dried sweet clover contains dicoumarol, which makes the moldy hay toxic for livestock by causing internal bleeding. The coumarin in sweet clover has a vanilla-like smell and is one of the base notes in the perfume industry's fougère olfactive category, which has a strong scent of herbaceous plants.

The plant has been used as a treatment for both internal and external inflammation, as a remedy for ulcers in the digestive tract, and as a lotion to reduce eye irritation. The Ramah Navajo gave cold sufferers a fragrant tea made from the leaves and also used it to wash the recuperating patient.

The Acoma and Laguna placed the fragrant plant in their bedding to repel bedbugs, while Great Plains Indians and early settlers put it in their sleeping quarters to freshen bedclothes and generally give the area a pleasant smell.

Ruminant wildlife and livestock forage on sweet clover, which provides high-quality nutrition, although palatability is reduced by the bitter taste of the coumarin. Small mammals and a wide variety of birds feed on the seeds, as well as the huge and assorted insect population attracted to sweet clover. This plant serves as a host for the larvae of eastern tailed-blue (*Cupido comyntas*), silvery blue (*Glaucopsyche lygdamus*), and Reakirt's blue (*Echinargus isola*) butterflies.

Sweet clover is popular for soil reclamation and stabilizing road cuts. It is also a nurse crop for revegetation projects, protecting small, slow-growing plants until they become well established.

FABACEAE (Bean Family)

Prosopis glandulosa

HONEY MESQUITE

The name "mesquite" comes to us from the Spanish *mezquite*, which in turn originated with the tree's name in Nahuatl (Aztec), *mizquitl*.

A fixture of the Southwest landscape, mesquite trees may grow forty feet tall, affording dappled shade with bipinnate leaves borne on graceful stems and trunks. The spikes of yellow flowers smell and taste sweet, attracting many species of bees that make a highly prized mesquite honey. Through these bees' pollinating services, mesquite produces long, straight pods, which are indehiscent, unlike other typical bean pods, which split open when ripe. The trees grow throughout New Mexico, except in the Four Corners area, in grasslands usually below 4,500 feet elevation.

The early residents of New Mexico found mesquite to be a reliable and nutritious food source even during drought years. As the pods ripen, the sugar content rises to more than 20 percent. Native Americans collected the pods in quantity and stored them in baskets to provide food throughout the year. The Acoma and Laguna ground the protein-packed beans into flour for breads and cakes, and they ate the beans raw or cooked like string beans. The sweet beverage the Apache extracted from the beans was sometimes allowed to ferment into a strong drink.

The Comanche chewed the leaves to settle an upset stomach. Some Native Americans used the sap as a remedy for sore throat, diarrhea,

Prosopis glandulosa, honey mesquite; field view.

wounds, and ulcers. They also dried the sap, ground it into a powder, and applied it to the navels of newborns to prevent infection.

Mesquite's hard, durable wood is used in various crafts for fashioning tools, furniture, flooring, and paneling, as well as for fence posts and charcoal. It is still sought as an excellent slow-burning fuel to use in smoking or barbecuing meat. The sticky sap was used as a glue to repair broken pottery. The inner bark provided fibers for weaving fabric, and the long, sharp thorns were used as sewing needles. The Pima put the dark mesquite resin in their hair to cover the gray.

Mesquite bean pods are good forage for wildlife and domestic livestock, although consumption by livestock is limited mainly to drought years. Many wild animals, including both birds and large and small

Prosopis glandulosa, honey mesquite; spines.

Prosopis glandulosa, honey mesquite; mature seedpods.

mammals, depend on the beans for food.

Those wise in the ways of nature have learned to keep an eye on mesquite trees (especially older ones) in the spring to determine when there is no longer any danger of frost. Mesquite bud break depends on a combination of day length, number of days with minimum temperature below 30°F, and soil temperature. When the local mesquites' computer system detects the optimal values, the leaves begin to emerge, and we know spring has officially arrived.

Mesquite is a facultative phreatophyte with an extraordinarily extensive root system that can penetrate forty feet deep, giving it access to water supplies that sustain growth even in zones with very low rainfall. Damage to the top part of the tree through drought, fire, herbicide use, and the like can be overcome by the ability of the crown to sprout new shoots. Mesquite of short stature should not be assumed to be young; a three-foot shrub may be a fifty-year-old survivor of environmental misfortunes. These survival skills have allowed mesquite to thwart the concerted and multipronged efforts at eradication, from herbicide spraying to chaining, that are regularly attempted throughout the Southwest.

Mesquite provides food for both caterpillars and adults of the leda ministreak butterfly (*Ministrymon leda*) and for the caterpillars of tailed orange (*Pyrisitia proterpia*), marine blue (*Leptotes marina*), ceraunus blue (*Hemiargus ceraunus*), Reakirt's blue (*Echinargus isola*), and Palmer's metalmark (*Apodemia palmeri*) butterflies.

FABACEAE (Bean Family)

Senna roemeriana (syn. *Cassia roemeriana*)

TWO-LEAF SENNA

Senna has linguistic connections with the Arabic and Hebrew words for "bush." *Senna* species in the Middle East were prescribed as laxatives as early as the ninth century and are still sold widely as over-the-counter cures for constipation. The species name honors Carl Ferdinand von Roemer, a nineteenth-century German geologist who is called "the Father of the Geology of Texas."

Two-leaf senna and its cousins, *Senna bauhinioides* and *S. lindheimeriana* (also pictured here), are found in southern and southeastern New Mexico, growing at lower elevations in dry sandy areas on rocky slopes and mesas. The plants flower in late summer to fall, forming curved seedpods that split on both sutures of the bean. *Senna lindheimeriana* is toxic to

Senna roemeriana, two-leaf senna; compound leaves with two leaflets and dehiscent (split) pods.

Senna bauhinioides, twinleaf senna.

Senna lindheimeriana, Lindheimer's senna; pinnately compound leaves.

cattle, causing serious muscle damage and gastrointestinal problems that lead rapidly to death.

Senna reseeds easily and is recommended for natural landscapes or xeriscapes with limestone soil. The plants are said to be moderately deer resistant. *Senna* is a host for larvae of the cloudless sulphur butterfly (*Phoebis sennae*).

FABACEAE (Bean Family)

Thermopsis montana

MOUNTAIN GOLDENBANNER, FALSE LUPINE, MOUNTAIN GOLDEN PEA

Thermopsis (hot appearing) apparently derives its name from its hot, bitter flavor. Or maybe we could say that the beautiful yellow flowers make this a hot-looking plant!

The common name "false lupine" was understandably bestowed because of the palmate leaves and lupine look-alike flowers—except that the leaves have only three leaflets, instead of lupine's five to nine, and the flower color is yellow, whereas lupine flowers range from white to blue and all shades in between but are never yellow.

Thermopsis montana, mountain goldenbanner; note typical hairy, vertical seedpods.

Goldenbanner is a montane and subalpine plant, as the species name indicates, growing in central, western, and northern New Mexico in ponderosa pine to spruce-fir zones from 6,000 to 11,000 feet elevation. The plants reach three feet tall, with long spikes of sunny flowers blooming from May through August. Masses of plants found together are the result of propagation by rhizomes.

The brilliant colors of the flowers and leaves, and the effect of the clustered growth habit, make goldenbanner a very attractive addition to natural landscapes.

There is a beast behind the beauty, however, as goldenbanner contains several very toxic alkaloids, including N-methylcytisine, anagyrine, and thermopsine, which cause profound muscle degeneration followed by death in cattle.

Goldenbanner hosts the caterpillars of western tailed-blue (*Cupido amyntula*) and Queen Alexandra's sulphur (*Colias alexandra*) butterflies.

Thermopsis montana, mountain goldenbanner; note trifoliolate leaves.

LOASACEAE (Stickleaf Family)

Mentzelia multiflora

MANYFLOWERED MENTZELIA, ADONIS BLAZING STAR, STICKLEAF

Christian Mentzel, for whom the genus is named, was a seventeenth-century German botanist, physician, and father of the first king of Prussia. As both the species and common names indicate, this plant is well adorned with flowers.

Mentzelia species have Velcro-like hairs on the underside of the leaves that attach tightly to clothes and animal hairs. Leaves that stick in sheep wool degrade its value.

Manyflowered mentzelia is widespread in New Mexico. The three-foot-tall plant is a biennial or perennial that blooms from April to August. It grows in grasslands or along roadsides in chaparral communities from 4,000 to 7,000 feet elevation. A close examination of the stamens reveals that the outer ones have a wide filament that is almost petal-like. The flowers open in the afternoon and close at sunset. The seedpods resemble a candle with a "wick" in the center; the wick is the remnant of the style leading to the ovary.

The oil-rich seeds were an important food source for the Ancestral Puebloans. On a kiva bench at Chaco Canyon, archaeologists discovered a jar with mentzelia seeds stored inside, and about three pounds of seeds were found in a pottery vessel in Cordova Cave in southwestern New Mexico. The Navajo still parch the seeds and grind them into flour.

Manyflowered mentzelia is a Life Medicine for the Navajo, who in the past used it to make a lotion for smallpox sores to keep them from pitting. They also used it for healing abdominal swelling, as an emetic,

Mentzelia multiflora, manyflowered mentzelia; whole plant.

and as an eyewash. The Keresan peoples considered it to be the strongest tuberculosis medicine available.

The Navajo made prayer sticks from mentzelia for the Beauty Way Ceremony. The Acoma and Laguna rubbed the leaves on their babies' thighs in hopes that they would become good, "stick-tight" horseback riders.

Mentzelia multiflora provides nectar for the silver-spotted skipper butterfly (*Epargyreus clarus*).

For information about another *Mentzelia* species, see p. 322.

ONAGRACEAE (Evening Primrose Family)

Calylophus hartwegii (syn. *Oenothera hartwegii*)

HARTWEG EVENING PRIMROSE, HARTWEG'S SUNDROPS

Calylophus hartwegii's genus name means "crested calyx," referring to the fingerlike projections on the tips of the sepals, which as a group are called

Calylophus lavandulifolius, lavenderleaf sundrops; note red stems.

a calyx. The species name honors Karl Hartweg, a nineteenth-century German botanist who made collecting trips to the New World and introduced many new species into cultivation.

This perennial is usually less than two feet tall and grows in sandy, clay, or caliche soils in grasslands and pinyon-juniper communities from 3,000 to 7,000 feet elevation. The flowers bloom from March to November, opening in the late afternoon and collapsing the next day. Hawk moths pollinate the flowers.

Hartweg evening primrose has a deep taproot and is very drought tolerant, making it an easy-care member of a xeriscape planting. Its showy yellow flowers attract a number of nectar-seeking insects.

The Navajo consider Hartweg evening primrose to be one of their Life Medicines.

Another yellow-flowered evening primrose is *C. lavandulifolius*, also pictured here, which grows at higher elevations than *C. hartwegii*, from 4,500 to 9,000 feet. Its soil requirements and value in landscaping are similar to those of Hartweg evening primrose. The Apache cooked the pods as a food for children.

For information about other members of the evening primrose family, see pp. 16, 113, and 328.

ONAGRACEAE (Evening Primrose Family)

Oenothera elata (syn. *Oenothera hookeri*)

HOOKER'S EVENING-PRIMROSE

Linnaeus created a word puzzle with the name *Oenothera*. Its etymology remains shrouded in confusion more than 250 years later. A strong case can be made for either of two alternatives. One is based on Pliny's assertion that it is "a plant whose juice in wine [*oeno* means "wine"] causes happiness," and the other asserts that the name means "donkey catcher," from *onos*, "donkey," and *thera*, "to hunt or to chase." Perhaps we can hybridize these to say that those who chase after wine make a donkey of themselves. The species name is Latin for "tall," an apropos adjective for this showy plant, which can reach up to six feet tall.

This biennial is just a low rosette of leaves in its first year and doesn't bloom until its second. It grows in moist soil in pinyon-juniper, ponderosa pine, and spruce-fir communities from 4,000 to 9,500 feet elevation. The leaves and red stems have long, sticky hairs. The flowers bloom from July to October, opening near sundown and emitting a pleasant fragrance to attract moths, which plumb the depths of the blooms with their long proboscises.

The small seeds of *Oenothera* species have a relatively high oil content, of which gamma-linolenic acid (GLA) is an important constituent. GLA has proved to be helpful in a number of conditions, including eczema, arthritis, premenstrual syndrome, cardiovascular disease, and diabetes.

One species, *Oenothera biennis*, is grown commercially for extraction of the oil and its GLA.

These beneficial properties have made *Oenothera* a Life Medicine for the Navajo, who use it in various ceremonies, to treat skin problems and colds, and to reduce arthritic swelling. The Paiute rubbed the root on the footwear of hunters to protect them from snakes and to bring the deer nearby. A yellow dye was also extracted from the plant.

Early settlers consumed the root when Irish potatoes were not available and used the plant to treat skin diseases, whooping cough, and asthma.

Oenothera elata is a statuesque feature in a natural landscape setting and does well in both sun and partial shade. Seed is commercially available.

For information about other members of the evening primrose family, see pp. 16, 111, and 328.

PAPAVERACEAE (Poppy Family) (formerly in Fumariaceae)

Corydalis aurea

SCRAMBLED EGGS, GOLDEN SMOKE, GOLDEN CORYDALIS

Corydalis aurea (golden crested lark) has an arched spur on the golden flower resembling the feathery crest of a lark. The common name "scrambled eggs" may sound tasty, but eating the flowers is not advised.

Corydalis aurea, scrambled eggs; seedpods and finely divided leaves.

This foot-tall plant grows throughout New Mexico in many types of soil, in both dry and moist conditions, and in several plant communities from 3,000 to 9,000 feet elevation. The flowers appear early in the spring and continue blooming to fall.

Corydalis aurea was used to control infections after childbirth, and Mexican herbal healers still recommend it as a postpartum drink to help new mothers regain their strength. The Navajo used it to stop diarrhea. When the Ojibwa were upset or anxious, they roasted the rhizomes over coals and inhaled the smoke for its sedative effect. The rhizomes of scrambled eggs have up to ten different alkaloids with anti-inflammatory properties, which the Navajo used to good effect in treating fever and pain, including arthritis pain.

The seeds are very attractive to ants, which harvest them for their elaiosomes—oil-bearing bodies attached to the seeds that are rich in protein. Ants routinely harvest seeds with elaiosomes and carry them to their nests, where they remove the elaiosomes and their nutrients and then toss the seeds into the rubbish heap, where they can germinate in a nutrient-rich environment. "Myrmecochory" is the name given to this symbiotic relationship between ants and plants. Elaiosomes may also act as a repellent to small mammals, protecting the seeds from excess consumption by animals.

ROSACEAE (Rose Family)

Potentilla fruticosa

SHRUBBY CINQUEFOIL, GOLDFINGER POTENTILLA, YELLOW ROSE

The genus name, *Potentilla*, comes from the Latin word for "potent" or "powerful," in reference to the many healing properties of species in this genus. The species name, *fruticosa*, means "shrubby."

Potentilla fruticosa has tiny palmate leaves with three to seven leaflets. Plants grow in moist meadows or along streams from 7,000 to 9,500 feet elevation, and the flowers bloom in June through August.

The Blackfoot of Montana used shrubby cinquefoil leaves as a spice, mixing them with dried meat. The Jemez used the beautiful yellow flowers in their summer dances. The Blackfoot filled their pillows with the leaves, and they used the dry, flaky bark as kindling. The Tanana of Alaska traditionally put the branches of shrubby cinquefoil under a girl's mattress when she reached puberty to lessen her first menstrual period and reduce the number of years of menstruation.

Animals graze the three- to five-foot-tall plants during the winter. For stockmen and wildlife biologists, *P. fruticosa* is as an indicator plant for overbrowsing, whether by big game or by livestock. *Potentilla fruticosa*, with its high tolerance of drought and poor growing conditions, is also useful for land reclamation and preventing erosion. As a landscape member, it is used as a hedge or border, and it is popular in urban areas because it needs little maintenance.

Also shown here are two closely related species, *P. gracilis* var. *pulcherimma* and *Pentaphyllum hippianum* (syn. *Potentilla hippiana*), which have similar flowers, growth habit, and medicinal value. *Potentilla gracilis* can

Potentilla gracilis var. *pulcherimma*, whiteleaf cinquefoil; note palmately compound leaves.

Pentaphyllum hippianum (syn. *Potentilla hippiana*), woolly cinquefoil; note pinnately compound leaves with green leaf surface and woolly underside.

be distinguished from *Pentaphyllum hippianum* by its palmate leaves (with seven leaflets), compared to the pinnate leaves of *P. hippianum.*

Pentaphyllum hippianum (common name "woolly cinquefoil") is a perennial that grows throughout New Mexico, except in eastern counties, in a number of plant communities from 4,300 to 12,000 feet elevation. It is usually part of a mature climax community but may also grow in early-successional communities.

A Life Medicine for the Navajo, woolly cinquefoil was used as an ointment for burns and as a poultice for wounds. Kayenta Navajo women took it during labor to speed the birth process. The whole plant is an astringent that constricts small blood vessels to help reduce bleeding, heal ulcers, and treat inflamed tissue.

SCROPHULARIACEAE (Figwort Family)

Verbascum thapsus

MULLEIN, FLANNEL MULLEIN, WOOLLY MULLEIN

This Old World émigré arrived in North America with the Puritans in 1630 and rapidly spread across the continent. The genus name is likely derived from *Barbascum,* an ancient Latin name that refers to its "hairy" leaves. The species name probably comes from the town of Thapsos in ancient Greece.

Mullein's six-foot spires are like long, vertical swords rising above rosettes of long, soft leaves that are completely covered with dense hairs, giving them a velvety feel. The flower stalk bolts in June through August of the second year. Mullein grows throughout New Mexico in disturbed soil in all plant communities from 5,000 to 8,000 feet elevation.

People always travel with their medicines, and when European settlers came to America, they brought with them seeds of mullein, a plant they evidently couldn't live without. As the settlers moved west, they

Verbascum thapsus, mullein; field view.

shared their home remedies with the native inhabitants, who not surprisingly adapted mullein to their own medicinal needs. Its reputation held it to be effective for many respiratory diseases, and thus it was called "lungwort" or "lung plant." People smoked the leaves or drank a tea made from the leaves to open bronchial passages when breathing was hampered by asthma or infection. Mullein was also an effective treatment for urinary tract infections, diarrhea, hemorrhoids, and migraine. Settlers even used it to cure ringworm, warts, and frostbite. Mullein leaf extract is touted as a treatment for a variety of disorders and is commercially available. Recent studies have affirmed the antibacterial and antitumor properties that have made this plant medicinally useful as far back as ancient Rome.

The large, velvety-soft leaves are sometimes called "cowboy's toilet paper," for obvious reasons, and they also make nice shoe liners for footsore hikers.

Verbascum thapsus, mullein; rosette stage.

Mullein is an interesting plant in a landscape, giving great vertical texture. It reseeds prolifically, and before you know it, there will soon be a veritable host of plants. Many insects visit the flowers, but only bees are able to accomplish actual pollination. The abundant seeds provide winter food for birds. Wildlife, including elk, deer, mountain goats, and some small mammals, feed on the leaves in winter.

SOLANACEAE (Nightshade Family)

Physalis cinerascens

YELLOW GROUND CHERRY, SMALLFLOWER GROUND CHERRY

Ground cherry has a little paper bag, the inflated calyx, around its green fruit, which resembles a "bladder" (*phys*). The leaves have an ashy-gray color (*cinerascens*) due to the dense covering of stellate trichomes.

This perennial is found in eastern New Mexico in dry sandy or rocky soil in grasslands and pinyon-juniper communities. The flowers bloom

Physalis cinerascens, yellow ground cherry; cutaway showing inflated calyx around young fruit.

from March through September and are campanulate (bell-shaped), with five fused petals bending their shy little faces to the ground. As the fruit matures, the calyx swells until it houses the pea-size fruit in a tiny balloon, as shown in the photo.

As a group, *Physalis* species are indicator plants of Ancestral Puebloan ruins. Archaeological evidence indicates that *Physalis* has been used for food for at least 1,100 years in the Southwest. Seeds have been found in both Bandelier National Monument and Salmon Ruins, with their presence in coprolites taken as strong proof of their inclusion in the diet of the local peoples. Modern Southwest Native Americans continue to use ground cherry, eating the fruits raw, cooked, or dried. Zuñi women mix the berries with chili, coriander, and onions to make a tasty salsa. Immature fruits, however, may be toxic.

Like its family mate *Solanum elaeagnifolium* (silverleaf nightshade; see p. 207), ground cherry's immature fruits and leaves contain the glycoalkaloid solanine, which has laxative and diuretic properties, among its other effects. Leaf teas are a treatment for urinary ailments such as nephritis and cystitis. Leaf extracts have shown antimicrobial action, including treatment of gonorrhea, as well as a salutary effect on cancer and diabetes.

ZYGOPHYLLACEAE (Caltrop Family)

Larrea tridentata

CREOSOTE BUSH, CHAPARRAL, GREASEWOOD

Creosote bush is the signature plant for the Chihuahuan Desert in central and southern New Mexico, as well as for the Sonoran and Mojave Deserts in Arizona and California, where it dominates great swaths of land in a monoculture.

Larrea tridentata, creosote bush; field view.

The genus name honors Don Juan Perez de Larrea, an eighteenth-century Spanish clergyman and patron of science. Understanding the significance of the species name requires very close inspection of the leaves, each of which is a pair of leaflets that are connate (fused) at the base, with a tiny bit of rachis sticking out between them, to give the *tridentata* (three teeth) effect.

Creosote bush is common on bajadas, dunes, and valley floors at elevations up to 5,000 feet. The six-foot-tall evergreen shrub has tough, waxy leaves that are sticky with resin and have a pleasant medicinal aroma. The golden flowers appear in the spring or anytime after a sufficient rainfall. The small fruits are white fuzz balls that are adapted for dispersal by tumbling along the ground while being propelled by the wind.

This plant mainly reproduces asexually by sprouting from the outside of the root crown, gradually expanding over the years into a giant network of cloned plants, which gives it the ability to be one of the world's oldest organisms. In California, the average life span of a clonal unit is estimated to be 600 to 1,200 years old, with one unit in the Mojave Desert said to be about 9,000 years old.

The resin of creosote bush has a high phenolic content, giving it both toxic and therapeutic properties. The allelopathic effects of these compounds in the leaves and roots effectively prevent any species except its own relatives from germinating and growing in its neighborhood. Medicinally, these phenols have antioxidant, antiseptic, and anti-inflammatory effects, but the dark side is that using too much of them or for too long can result in kidney and liver damage. *Larrea* was included in the *U.S. Pharmacopoeia* during the second half of the 1800s because of its antiseptic qualities, and it was used as an FDA-approved food preservative in the mid-1900s, but both listings were withdrawn in light of toxicity reports.

The value of the plant was not news to Native Americans, who had been using its treasure chest of cures for centuries. The Papago long ago realized that rheumatism responded well to a warm poultice of heated creosote branches. They also sprinkled dried, powdered leaves as a disinfectant on a newborn's navel to promote healing and even used this powder as a body deodorant. They used the leaves, after chewing them, to reduce swelling and prevent infection due to bites from insects, spiders, scorpions, and snakes. The phenolic content of the plant made it effective for respiratory congestion. Native American women relied on the plant to help deal with the pain of menstrual cramps and childbirth and to promote milk flow after childbirth.

Creosote bush has an ecologically and economically useful relationship with a tiny scale insect, *Tachardiella larreae*, which lives in tight little colonies along the stems. The insects extract sugars to manufacture a resin called lac, which they excrete onto the stems. Some of the medicinal properties of creosote are actually a combination of lac and the plant. Infusions of leaves and lac were used to treat skin problems, such as cuts, abrasions, and boils. The lac was used by various tribes as an all-purpose glue and sealant. When lac is heated, it becomes soft and pliable, then hardens upon cooling to form an impervious barrier that was very useful for mending pottery or waterproofing baskets.

Creosote branches were used as roof thatch for huts where menstruating women temporarily lived. The bush supplied wood for foreshafts of arrows and handles for gourd rattles. Charcoal from burned creosote was used as a permanent greenish-blue tattoo pigment.

As a feature in a xeriscape or natural landscape in the Southwest, creosote bush is useful as an individual or a hedge, providing not only color and texture but also a unique olfactory experience.

ZYGOPHYLLACEAE (Caltrop Family)

Tribulus terrestris

GOATHEAD, PUNCTUREVINE

Puncturevine is a cleverly designed antipersonnel device guaranteed to puncture bare feet and bicycle tires. Both the common family name, "caltrop," and the genus name, *Tribulus* (a three-pronged caltrop), point to the resemblance of this plant's multipronged seedpods to a military weapon. Caltrops, which are iron balls with projections that were scattered about to damage the enemy's feet, were used by Roman soldiers more than two thousand years ago. Caltrops were ingeniously designed so that they were always in the "on" position, with at least one of the three or more prongs pointing upward into the path of the enemy's feet and the other prongs stabilizing the weapons in the earth. The species name, *terrestris*, recognizes the "earth"-hugging growth habit of puncturevine.

Tribulis terrestris, goathead; aerial view of vine.

Tribulis terrestris, goathead; seedpod made of three to five spiny nutlets.

This native of the Mediterranean spread throughout North America beginning in the 1800s, probably along railway lines. As shown in the photo, this deep-taprooted annual vine forms mats of two to three feet in diameter in disturbed soil from 3,500 to 7,000 feet elevation. The "stickers" have strong spines and when mature break into three to five nutlets. The roots have an association with nitrogen-fixing bacteria.

Goathead has played a role in herbal medicine in the Old World, where the plant originated, but it has seen little use among Native Americans. The Navajo, however, smoke "bull head," as they call it, in the Evil Way and Bead Way Ceremonies. They also use it for blackening their skin in the Evil Way Ceremony.

In Europe, *Tribulus terrestris* has been a traditional treatment for constipation, sexual dysfunction, and nervous disorders. In China, it has been valued to treat liver, kidney, urinary tract, and cardiovascular maladies. In India, the plant has been used as an aphrodisiac, a diuretic, and a medicine for calming the nerves.

Goathead contains steroidal saponins, on the basis of which it has been promoted for bodybuilding and improving male sexual performance. While controlled studies have failed to conclusively demonstrate significant improvement in either muscle mass or testosterone levels in humans, the bank accounts of retailers are undoubtedly much more buff.

Goathead, or possibly a fungal toxin associated with it, can cause photosensitivity in grazing sheep, causing the skin to peel, with the resulting loss of lips and ears, and damage to the liver. It can also accumulate enough nitrates in drought periods to be toxic to sheep. Nitrates compete with oxygen in binding to red blood cells and, in effect, cause suffocation.

Green Flowers

AMARANTHACEAE (Amaranth Family)

Amaranthus palmeri

PIGWEED, CARELESS WEED

Several species of *Amaranthus* are well known for the beautiful reddish color of their inflorescence. This rich color does not fade as the plant matures, suggesting "immortality," as connoted by the Greek word *amarantos*. The much-maligned pigweed received its common name because of the great fondness that pigs (and other livestock) have for this highly nutritious weed. The species name recognizes the Englishman Edward Palmer, who in the late nineteenth century immigrated to America and became a very productive collector of plants and artifacts in the Southwest. It is often difficult to distinguish one *Amaranthus* species from another because there is so much hybridization among them, especially the weedy ones.

Unusual among plants, the protein of amaranth is complete—that is, it has a well-balanced amino acid profile. The amino acid lysine, which in corn and other cereals is notably deficient, is proportionally high relative to other amino acids in amaranth.

In New Mexico, *Amaranthus* grows in disturbed soil in agricultural regions and blooms from summer through fall. It is not favorably regarded by farmers because it is such a prolific weed and has become difficult to control in cultivated land. Native Americans, however, had a high regard for it. Ancient Puebloans ate it as a green vegetable, boiled, fried, or cooked together with green chili and meat. The Acoma parboiled young leaves and dried them for future use.

The tiny, shiny, black amaranth seeds were also important as a food source. The Apache and Ramah Navajo ground the seeds into flour for bread. The seeds also can be used to make porridge cooked with goat milk or parched to eat as a snack.

The Navajo used amaranth leaves mixed with those of other plants for ceremonial smoking in the Coyote Chant. Amaranth was combined with *Artemisia* species (pp. 139–142) to make a liniment for the Bead Chant.

Amaranthus, like other members of its family, contains rich reddish pigments called betalains, which give the characteristic color to beets, red chard, and ornamental *Amaranthus* flowers. These betalain pigments are found primarily in amaranths and cacti and differ from the more common red anthocyanin pigments found in other plant families. Amaranth pigments were once commercially extracted as FD&C Red Dye #2 but subsequently were replaced by synthetic dyes. The Hopi and Zuñi used amaranth flowers to give a red color to ordinary white or yellow cornmeal when making bread for ceremonial dances. The Navajo crushed the leaves and flowers to obtain a reddish pigment used for rouge and for dyeing wool.

Pigweed can be toxic to ruminants if the soil has a high nitrate content, especially during drought, because amaranth will accumulate nitrates in its leaves. Nitrates in the blood compete for oxygen, effectively suffocating the victim. Pigweed also can cause abortion and bloat in ruminants.

Pigweed is the food source for caterpillars of the common sootywing butterfly (*Pholisora catullus*).

AMARANTHACEAE (Amaranth Family)

(formerly in Chenopodiaceae)

Atriplex canescens

FOUR-WING SALTBUSH, CHAMISO, WHITE GREASEWOOD

Four-wing saltbush was "born" for a life of both adversity and service. It thrives in the most inhospitable environments, performing well in drought, heat, and saline soil, and serves as a generous source of food and medicine not only for humans but also for wildlife and livestock.

The genus name, *Atriplex*, comes to us from Pliny, who two thousand years ago gave this name to orach, *Atriplex hortensis*, the European kitchen vegetable that was widely popular until spinach became available. The species name, *canescens* (turning to gray), describes the color of the leaves, which are typically grayish due to salt deposits, called "scurf," that look like dandruff. Leaves of four-wing saltbush are covered with tiny flattened hairs (trichomes) that act as a botanical desalinization factory when the plants are growing in salty soil, concentrating and removing the salt from the water taken in from the root environment. The salt is deposited onto the leaf surfaces out of the way of the plant's cellular metabolism. The plant's common name is derived from the four wings around the covering of the small seeds (see photo).

Four-wing saltbush grows throughout New Mexico at elevations up to about 8,000 feet, blooming from April through October. It is usually dioecious, with female flowers on one plant (see main photo) and males (see photo) on another. But any individual plant doesn't have to remain the same sex forever! Four-wing saltbush shakes things up by turning from a female into a male after an especially harsh winter or severe drought.

Atriplex canescens, four-wing saltbush; male flowers.

American Indians have long used four-wing saltbush as a food source, as evidenced by the presence of seeds in feces at archaeological sites. It is also an indicator plant for ancient ruin mounds because it grows well in the rich soil of middens, or ancient family garbage dumps. Springtime-tender leaves were cooked as greens and eaten fresh as a snack food. The Ramah Navajo included the older leaves and stems in the roasting pit with corn to give it a salty taste. The Navajo used the flowers to make a pudding. The Paiute and Navajo parched the seeds and ground them into flour to make bread or porridge.

Using the same method of nixtamalization as Mesoamerican Indians, the Navajo and other corn-producing Indians in North America processed their corn with alkaline ashes from four-wing saltbush. This process not only softens the hull of the corn to allow for easy removal but also makes the various seed components more nutritionally accessible. Increased availability of the amino acids lysine and tryptophan, along with niacin (vitamin B_3), is among the more notable benefits of nixtamalization. Both protein deficiency and pellagra became common

Atriplex canescens, four-wing saltbush; seeds with wings.

when corn was introduced around the world following the voyages of Old World explorers to America because knowledge of the nixtamalization process did not accompany its spread.

Saltbush ashes were also added to finely ground meal made from blue corn and mixed with water to make piki bread, an important food for the Pueblo peoples. The alkaline ashes maintained the blue color of the cornmeal as the wafer-thin sheets were cooked on a stone griddle.

The Jemez made a poultice from the leaves or flowers to treat ant bites. The Navajo made a poultice of the roots to treat toothache. They used the leaves and roots for respiratory ailments, making a snuff from the leaves to treat nasal problems and a strong tea from the leaves and roots to alleviate coughs.

Saltbush was among the important dye materials for the Navajo, who mixed the leaves and flowers with alum to make a yellow dye for wool. They also added saltbush ashes to intensify the red dye made from mountain mahogany (see p. 332), which was used for coloring buckskin.

Saltbush leaves were used as soap for washing hair and cleansing itchy rashes, such as those from measles and chicken pox. The Zuñi mixed the flowers with water to make a hand soap.

As part of their manhood rites, young Navajo men were given arrows made from four-wing saltbush wood (see photo). The ceremonial counterpart for young women involved the presentation of "female weapons," or stirring sticks, made of greasewood, as discussed on p. 255.

Throughout the western United States, four-wing saltbush is a valuable forage shrub for livestock and wildlife, including ruminants, small mammals, and birds. In winter, it becomes especially valuable because its protein content (up to 18 percent) and carotene levels rise during the cold months.

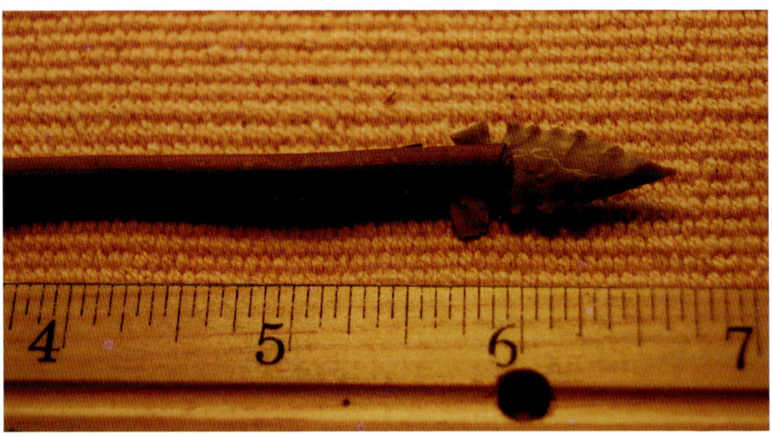

Example of arrow shaft. Artifact found in cave on Lower Pecos River.

Atriplex obovata, New Mexico saltbush.

Saltbush exhibits wide adaptability across all habitat zones and tolerates harsh environmental conditions, making it a valuable tool for erosion control or reclamation of disturbed or degraded land, such as mining sites or oil exploration and drilling sites. The USDA has released several improved cultivars, including Marana, Rincon, and Santa Rita, that are well adapted to New Mexico.

Four-wing saltbush hosts the caterpillars of western pygmy blue (*Brephidium exilis*), Mohave sootywing (*Hesperopsis libya*), and saltbush sootywing (*Hesperopsis alpheus*) butterflies.

The short (one-foot-tall), shrubby *Atriplex obovata* (see photo) is found in western New Mexico, growing in sandy soil and providing desirable browse and forage for wildlife and livestock, especially in the spring, when fresh leaves emerge. It also is an important soil stabilizer.

AMARANTHACEAE (Amaranth Family)

(formerly in Chenopodiaceae)

Chenopodium berlandieri

NETSEED LAMBSQUARTERS, PITSEED GOOSEFOOT, SOUTHERN HUAUZONTLE

Chenopodium's large leaves are shaped like "goose feet," which is what the name means in Greek. The French botanist and explorer Jean-Louis Berlandier, who made extensive botanical collections in Mexico and Texas in the early 1800s, is honored with the species name.

For centuries, netseed has been cultivated as a food crop in the highlands of central Mexico, and through trade among Native Americans it entered North America. Regionally distinct varieties were selected for leaf or for seed production.

Netseed grows throughout New Mexico in disturbed soil, as do most domesticated crops, and over time has interbred with various local weedy *Chenopodium* species. The plants are tall, reaching up to eight feet in good growing conditions. The inflorescence is a loose, drooping panicle, or cluster of flowers. Flowering is late, usually August through October, and seeds mature just before frost.

Many Native American tribes, including the Navajo, Papago, and Zuñi, have used both the leaves and seeds for nutritious, high-protein food. The leaves are harvested from spring through fall for frying or boiling like spinach and for drying for later use.

The leaves have a glandular surface that is strongly pigmented with betalain (see photo), a reddish-purple dye that is characteristic of beets and other amaranth family members, as well as cacti, and that differs from the more typical anthocyanin pigments found in other plant

Chenopodium berlandieri, netseed lambsquarters; young inflorescence showing betalain pigment.

species. Betalain dyes are useful as food coloring, and because they are also natural antioxidants, they are of increasing interest to the food industry.

The tiny, shiny black seeds contain saponins, which are removed by soaking overnight and rinsing and then broken down further by cooking. These saponins are not harmful and pass through the system with no further effects. Netseed leaves contain oxalic acid, which aggravates gout, rheumatism, and arthritis and causes kidney stone formation.

Netseed is a host for the larvae of several butterflies, including the Mexican sootywing (*Pholisora mejicanus*) and Mohave sootywing (*Hesperopsis libya*).

AMARANTHACEAE (Amaranth Family)

(formerly in Chenopodiaceae)

Krascheninnikovia lanata (syn. *Eurotia lanata*)

WINTERFAT

As its species name, *lanata*, indicates, winterfat is a "woolly" plant, covered with long, fine hairs. The genus name was given in honor of a Russian botany professor and explorer who traveled extensively in Siberia in the early 1700s, writing reports of the natural history of the region. The common name, "winterfat," refers to its importance as a winter browse for sheep and wild ruminants. Even its Navajo name, meaning "jackrabbit food," acknowledges its importance for all wildlife.

Winterfat is found throughout New Mexico, except in the southeast, in creosote bush scrub and pinyon-juniper woodlands from 3,500 to 7,000 feet elevation. The plant is dioecious, with male and female (see photo) reproductive parts on separate plants. The flowers bloom from May to July. The fruit is enclosed in a two-beaked tube formed from bracts that earlier surrounded the flower.

Pueblo groups used winterfat to treat skin problems. To relieve poison ivy pain, they chewed winterfat leaves to make a poultice and applied it to the rash. They also used the leaves to treat boils and related skin sores, including smallpox pustules. The Tewa and Zuñi used the powdered roots to treat burns. The Tewa decocted the leaves to reduce fever, and the Navajo used a similar decoction for patients spitting up blood. For overdoses of *Datura* (see p. 339), the Ramah Navajo drank winterfat tea until the effects abated.

Krascheninnikovia lanata, winterfat; close-up of female flower with long hairs.

In a xeriscape, winterfat provides interesting visual texture, with its hairy, willowy stems rising from the woody base. It is heat, drought, and cold tolerant, and in fact needs cold nights to thrive. Plants grow quickly and become very bushy in response to heavy pruning in early spring.

ASTERACEAE (Sunflower Family)

Ambrosia artemisiifolia

COMMON RAGWEED, ANNUAL RAGWEED

In spite of its genus name, which means "food for the gods," *Ambrosia artemisiifolia* had only medicinal uses among Native Americans. The species name simply means it has leaves like *Artemisia* (see p. 139).

Ragweed is classified as an invasive annual weed because it spreads so aggressively in agricultural areas and in other disturbed sites. The plant blooms from July to October. Ragweed is monoecious, with the male flowers at the top of the long inflorescence and the female flowers below.

Ambrosia acanthicarpa, flatspine bur ragweed; inflorescence.

The male flowers produce prolific quantities of pollen that is responsible for much suffering by people afflicted with allergies. The mature achenes are hard and prickly, but they have a high oil content and are an object of nutritional interest to birds.

Ragweed's historical association with the Ancestral Puebloans is indicated by the presence of ragweed pollen inside the ruins of dwellings at Aztec Ruins National Monument. The Cherokee used ragweed leaves to sooth insect bites and hives, and the Mahuna used it to treat diseases of the scalp. The Dakota made a tea from the leaves and flowers to treat bloody diarrhea and vomiting. The Iroquois made a tea from the roots to treat stroke, and the Houma used root tea for menstrual problems. Leaf tea also addressed skin infections and stroke.

A cousin of common ragweed, *A. acanthicarpa* (flatspine bur ragweed, annual bur-sage) has female structures with long, flat spines (see photo) that are the inspiration for one of the common names, as well as the species name, *acanthicarpa*, meaning "spiny seed." This late-fall-blooming weed grows throughout New Mexico, except in the southeast, in disturbed areas up to about 9,000 feet elevation. It is remarkable for its robustly spiny, pink to deep purple female inflorescence borne below the spike of male flowers. The leaves are deeply pinnately divided, giving them a frilly look. The petiole, which attaches the leaf to the stem, has "wings."

Zuñi women drank a tea made from the whole plant to reactivate the menstrual cycle, effectively serving as an abortion agent. The Zuñi also ground the roots and used the powder to treat toothache. The Ramah Navajo burned the leaves and used the ashes as a skin-blackening agent in the Evil Way Ceremony.

ASTERACEAE (Sunflower Family)

Artemisia tridentata

BIG SAGEBRUSH, NAVAJO SAGE

Big sagebrush is one of the most important shrubs in the western United States, both in the extent of its distribution and in its economic value. It is the climax species in almost all of its range and occurs in communities with four-wing saltbush, black greasewood, pinyon pine, Utah juniper, and ponderosa pine. In New Mexico, it is found throughout the northwest and central regions. Early settlers in the American West recognized that big sagebrush grew on land very suitable for agriculture, and they established farms where big sagebrush grew near a surface water supply.

Big sagebrush is a perennial evergreen growing in deep, rich soil and is normally three to ten feet tall, depending on the moisture gradient. The "three teeth" decorating the end of its leaves (see photo) are recognized in the species name, *tridentata*. The roots extend deep and wide, with hydraulic lift provided by the taproot, which brings water near the

Artemisia tridentata, big sagebrush; note three lobes on end of each leaf.

surface during the day and then releases it for uptake by the shallow root system. Plants typically are forty to fifty years old, with some reaching one hundred years.

The defining feature of *Artemisia* species is their medicinal fragrance of camphor and eucalyptol. Big sagebrush is believed to contain an even greater concentration of these and other aromatic oils than other *Artemisia* species and is therefore preferred over the others for medicine. Big sagebrush is one of the Navajo Life Medicines. The Navajo inhaled the scent of the aromatic oils from the leaves and stems as a decongestant and to reduce coughing. Navajo midwives used it to relieve postpartum pain. The Zuñi put the leaves in shoes to treat athlete's foot and to deodorize the feet. The Kayenta Navajo used it to treat wounds on sheep. Big sagebrush has antihelminthic properties used to expel worms, and early settlers treated Rocky Mountain spotted fever with big sagebrush tea.

A root tea was rubbed on animal hides during the tanning process, and the easily shredded bark was used to make woven skirts, leggings, shirts, ropes, and baskets. The Navajo use big sagebrush in the Night Way and Water Way Ceremonies. It is also a source of yellow-green and gold dyes.

Ruminant wildlife and livestock heavily browse big sagebrush during the winter months, when more preferred food is not available. It has a high protein content and is easily digestible except where terpenoid levels are high. It also provides food and cover for birds, especially sage

Artemisia dracunculus, false tarragon; inflorescence.

Artemisia dracunculus, false tarragon; note leaf galls.

grouse, and for small mammals. All animals show varying preference for big sagebrush according to the subspecies, local populations, and even individual plants, as the chemical profile is influenced by location, type of soil, and the ploidy level, or number of chromosomes.

Big sagebrush is easy to establish by seed, and because it has such broad adaptability in different regions, it has become important for revegetating damaged mining sites.

It hosts the caterpillars of Old World swallowtail (*Papilio machaon*) and American lady (*Vanessa virginiensis*) butterflies and the Hera buckmoth (*Hemileuca hera*).

Two other *Artemisia* species that share similar chemical properties are *Artemisia dracunculus* and *A. filifolia* (see photos). *Artemisia dracunculus* grows throughout New Mexico and is generally found in sandy soil in disturbed areas, in grasslands, or on the edge of woodlands. It has reddish stems and often has leaf galls. It was important among the Navajo for healing wounds.

Artemisia filifolia, sand sagebrush; field view.

Artemisia filifolia grows throughout most of New Mexico, growing in sandy soil at elevations up to 6,500 feet. It is the dominant plant in the communities where it is found. Its tiny flowers bloom from August to September.

ASTERACEAE (Sunflower Family)

Conyza canadensis

HORSEWEED, MULETAIL, BLOODSTANCH

Horseweed will never win a beauty contest, but this potentially invasive weed has a fascinating chemical portfolio that makes it worthy of attention. It is found throughout most of New Mexico, and indeed most of North America.

More than 2,300 years ago, Aristotle gave the Latin name *conyza*, or "flea," to some plants in the sunflower family, and it was officially co-opted as the genus name about two hundred years ago. The species name means "from Canada."

When the plant is flowering, an aromatic oil can be extracted for medicine as well as for flavoring food. Horseweed oil was listed in the *U.S. Pharmacopoeia* in the 1800s and continues to be used by herbalists who practice homeopathic medicine. The oil is also commercially available as an FDA-approved food flavoring agent in baked goods, beverages, candy, ice cream, and sauces, where it imparts a fruity or spicy flavor.

The Navajo applied a hot poultice of horseweed to babies with a postnatal infection. They included it in a lotion used for pimples and snakebite. The Acoma and Laguna used it to lighten skin blotches and dark spots. Others made a tea of the leaves and roots to treat digestive tract disorders, including diarrhea, stomach pain, hemorrhoids, and meteorism—an astronomical term that vividly describes the affliction of constantly passing "wind." It was applied as a poultice to aching muscles to reduce inflammation and pain. A hot tea was used to relieve cold symptoms such as stuffy head, sore throat, and cough. As one of its other common names, "bloodstanch," indicates, it is effective in stopping bleeding. Horseweed is also a diuretic and is used to treat urinary tract infections.

During a drought when forage is limited, livestock will eat young horseweed plants, even though it's not on their list of favorites. The unknown toxins damage the brain, causing the animal to walk in circles, become excessively excitable, tremble, and finally die. If the symptoms are recognized in time, the animal can be revived with the B vitamin thiamine.

ASTERACEAE (Sunflower Family)

Xanthium strumarium

COMMON COCKLEBUR

Common cocklebur is indigenous to the New World and is one of America's earliest and least desirable exports, first appearing in Europe half a century after Columbus arrived in America.

This five-foot-tall annual is typically found in sandy or clay soil in moist areas around playa or saline lakes and irrigated fields. It blooms from midsummer through the fall. The flowers are monoecious, with males and females in separate locations: the male flowers are in clusters at the top, and the females (the burs) grow below in the axils of the leaves. The female flowers can also create seeds apomictically, without the need for pollination. The burs are a particular nuisance when they attach to the fur of animals such as sheep, and they sharply degrade the value of the wool.

In each seedpod, there are two seeds, the largest of which germinates first. The second seed is an insurance policy, becoming dormant to await future germination. The seeds themselves and the cotyledons, or first leaves that appear after germination, are extremely toxic to all livestock. The agent of death is the glycoside carboxyatractyloside, which causes hypoglycemia (low blood sugar) and destruction of the liver. The toxin rapidly diminishes as the first true leaves emerge. Unfortunately, cocklebur seedlings are very palatable to cattle, sheep, and swine, and death is not uncommon among animals that browse the young plants emerging after a rain or when water recedes from flooded areas.

Xanthium spinosum, spiny cocklebur.

Cocklebur has an arsenal of other chemicals, including sesquiterpene lactones, which have insecticidal properties to inhibit insects from feeding on the plant and allelopathic properties to prevent seeds of other plants from germinating in its neighborhood.

Cocklebur also exhibits antimicrobial activity. The Acoma and Laguna spread cocklebur seed powder on skin sores to promote healing, both for themselves and for their horses. Others treated wounds with a tea made from the whole plant or a poultice of crushed leaves and burs. The Cherokee used the plant as an emetic, to treat cramps, and to relieve croup. The Pima of Arizona used cocklebur pulp to soothe sore eyes. The Navajo used it as an antiperspirant.

The Acoma and Laguna ground the seeds to a powder used to make a blue paint for masked dancers. The Ramah Navajo still use cocklebur leaf ashes for ceremonial skin blackening.

Birds and butterflies are unaffected by cocklebur toxins. Mourning doves eat cocklebur seeds to some extent, and larvae of the bordered patch butterfly (*Chlosyne lacinia*) eat the leaves.

A related species, *Xanthium spinosum*, arrived in the Southwest from tropical America, probably with Spanish settlers. It is easily distinguished by the robust forked spines at the leaf bases, as shown in the photo.

CUPRESSACEAE (Cypress Family)

Juniperus spp.

JUNIPER

The name *Juniper* is derived from the Latin name *gigiperus*, given by Pliny and Virgil two thousand years ago. Its value and reverence in Europe through the centuries are identical to its importance in the American Southwest. *Juniper* species are very significant sources of food, medicine, timber, and firewood for Native Americans, and they provide food and cover for many wildlife species. The various species share many growth and habitat characteristics as well as end uses. Here we provide distinguishing information for three key species and then describe their many similar uses in daily life in the past as well as the present.

Junipers grow on dry rocky slopes from 4,000 to 7,000 feet elevation, often in association with pinyon pine. They are gymnosperms, reproducing by cones rather than flowers.

***Juniperus osteosperma*, Utah Juniper:** This is a climax species found mainly in the Four Corners area of New Mexico. Trees reach ten to twenty-five feet tall and may live up to 650 years. The tree is monoecious, bearing both female (see main photo) and male (see photo) structures. The female cones contain one or two seeds and are dry and mealy when crushed. The bark typically shreds in long, thin gray-brown strips, as shown in the photo. The seeds are not produced until the tree reaches about thirty years old. This species grows too slowly to be of any value in land rehabilitation projects. It has the highest volatile oil content of the three species described here. These volatile oils inhibit rumen microbes; deer, which are ruminants, can scent these volatile oils and avoid eating it.

Juniperus osteosperma, Utah juniper; field view.

Juniperus osteosperma, Utah juniper; staminate (male) cones.

Juniperus osteosperma, Utah juniper; shreddy bark.

***Juniperus deppeana*, Alligator Juniper:** This is one of the largest junipers, reaching twenty to forty feet at maturity. The single short, massive trunk is deeply furrowed into checkered plates like an alligator's back (see photo). The tree is dioecious, with some trees having male cones and others female cones. The leaves, which are flatter than those of other species, are more palatable and have less volatile oil content, making them more heavily browsed by ruminants such as deer. The berrylike cones, an important food source, contain three to five seeds and remain on the tree until they mature in the second year. Easily propagated by seed, alligator juniper can also sprout prolifically from the roots or crown, making it useful for land rehabilitation and erosion prevention. This species is found mainly in southern and western New Mexico.

Juniperus deppeana, alligator juniper; note heavily scaled bark.

***Juniperus monosperma,* One-seed Juniper:** This small, symmetrical tree is usually about ten feet tall, with many stems. The bark shreds into long strips. The taproot can reach down to twelve feet. The leaves are browsed by ruminants such as deer. The tree is dioecious, with female trees producing succulent cones containing one seed (see photo) and maturing in one year, providing abundant year-round food for many wildlife species. One-seed juniper is useful for long-term land rehabilitation projects on coal mine and uranium mine land. It is found throughout New Mexico.

Juniperus monosperma, one-seed juniper; note cone and single seed.

Juniper timber used in construction, with shredded bark for insulation. Photo taken at Salmon Ruins.

At Salmon Ruins, archaeologists found eight burial sites containing juniper seeds to sustain the departed on their journey, offering evidence that juniper has long been a source of food for New Mexico's Native American populations. Modern groups of Native Americans continue to use juniper for food. The Acoma flavor meat with the berrylike cones, and the Tewa eat the cones with piki bread. In the fall, the Navajo add the red tips of the leafy branches to cornmeal to add flavor and color. Juniper cones have been shown to help regulate blood sugar, a healthy feature for diabetics.

Juniper cones are added for flavor to a variety of foods and beverages, as well as to some types of tobacco. Juniper provides the unique taste of the alcoholic beverage gin.

Many uses of juniper revolve around the bearing of children. Zuñi women drank a juniper tea as a contraceptive, and Navajo women took it to relieve menstrual pain. Tewa women who wanted a female baby drank a decoction of the leaves. Ramah Navajo women with difficult labor inhaled juniper smoke to facilitate the birthing process. The Tewa used juniper to make a postpartum wash for mothers and rubbed juniper ashes on newborns. The absorbent bark was used to make sanitary napkins for women and disposable diapers for babies.

Juniper also has antiseptic properties. It was applied to sores and wounds to prevent infection and was placed inside decaying teeth. Juniper leaf tea is a stimulant and a diuretic; a very strong tea has an emetic effect. The Acoma and Laguna used the pollen cones to stop diarrhea and settle an upset stomach, and they chewed the bark for a laxative. Traditional healers still sell juniper extract, which they call "cedar oil," as a treatment for skin problems, aches and pains, nervous tension, colic, flatulence, and lung disorders.

Since juniper wood resists termite damage and fungal decay, it is often used for mine timbers, fence posts, shingles, coffins, and pencils. In the past, the easily shredded bark was used to insulate buildings (see photo) and was stuffed into moccasins and leggings for warmth in the winter. It was a source of fibers from which mats, clothes, blankets, pot rests, and saddles were fashioned. Bark-lined seed storage bins helped keep the contents dry, insulated, and free of mold. The soft bark was also used as toilet paper, and today as in the past, young boys find the fibrous bark a flavorful tobacco substitute. Juniper sap serves as an adhesive.

The Tewa carried juniper seeds in leather pouches to protect against evil. The Navajo still string the seeds into necklaces for children to wear to keep nightmares away.

All junipers serve as a larval host for the juniper hairstreak butterfly (*Callophrys gryneus*).

EUPHORBIACEAE (Spurge Family)

Croton texensis

TEXAS CROTON, DOVEWEED, SKUNKWEED

Croton seeds are round and spotted, just like an engorged tick, which is what the genus name means in Greek. The species name, *texensis*, reveals its connection with New Mexico's neighbor to the east.

Texas croton grows throughout New Mexico in disturbed sandy soil in a variety of plant communities up to about 8,000 feet elevation. The plants are silvery, thanks to their covering of white, stellate hairs, and are delicately branched, with small flowers at the ends. A close examination reveals that they have the typical spurge family anatomy: no petals, but a large ovary spilling out of the female flowers, and the stamens well exserted in the male flowers. The plant is usually dioecious, meaning that male and female flowers are found on different plants. Flowering begins in late spring and continues through the fall.

A similar species, *Croton pottsii,* is found mainly in southern New Mexico, where it overlaps the range of *C. texensis*. The two can be distinguished by *C. pottsii*'s wide, ovate leaves and slightly more compact shape, compared to *C. texensis*'s narrow, linear leaves and more open, or airy, growth habit.

While Texas croton as a plant is nondescript, its chemistry packs quite a punch. The oil from the seeds produces an extreme purgative response, which many tribes throughout New Mexico made use of as a laxative. The Acoma and Laguna also put the ground seeds into open sores to promote healing, and they made a salve with the crushed roots to use as a hemorrhoid treatment. Their medicine men would chew croton root prior to sucking out a snakebite and then make a poultice to apply to the wound. The Zuñi also treated gonorrhea and syphilis with croton leaf tea.

The Navajo burned croton to smoke clothes that had been sprayed by a skunk. Croton is also a good insect repellent, and several tribes burned the powdered leaves to chase away unwanted bugs. In spite of these insecticidal properties, Texas croton is a larval host and nectar source for the goatweed leafwing butterfly (*Anaea andria*).

Croton's seeds are a moderately good food source for terrestrial birds such as doves and quail. Grazing animals avoid eating croton leaves even in the absence of other food, and thus it can become invasive in poorly managed rangeland.

PLANTAGINACEAE (Plantain Family)

Plantago patagonica

WOOLLY PLANTAIN, WOOLLY INDIANWHEAT

Woolly plantain will never win an award for beauty, but its many medicinal uses are worth making its acquaintance. The genus name, *Plantago* (footprint), refers to the wide leaves of other species in this genus, which resemble a foot. *Plantago patagonica* was originally described from plants in Patagonia, a region in Chile and Argentina.

All parts of the plant, except the membranous tan petals, are woolly with hair. This short, easily overlooked annual grows in all soil types, especially rocky or sandy soil, in disturbed areas of desert and grassland zones from 2,800 to 6,000 feet elevation. The flowers bloom early, from February to July.

Plantago patagonica and two other species, *P. helleri* and *P. rhodosperma* (see photo), are native species. *Plantago lanceolata* (see photo) was introduced from Europe. Aside from *P. patagonica*, the other species have limited distribution in New Mexico.

The mild-tasting, nutritious leaves may be eaten as a cooked vegetable or as a raw salad ingredient. The Kayenta Navajo made the seeds into a mush. The Navajo drank plantain tea to reduce their appetite, a benefit on long treks.

Plantago rhodospermum, redseed plantain, and *P. helleri*, Heller's plantain (right side).

Plantain leaves and roots contain hemostatic enzymes, which constrict blood vessels, making the plant valuable in treating cuts, wounds, and abrasions. Its demulcent and anti-inflammatory properties made it useful as a poultice for rashes, stings, burns, irritated eyes, and hemorrhoids. A root infusion to stop diarrhea was gentle enough for a baby. A diuretic tea brewed from the leaves was used to treat urinary tract infections and chronic lung problems. Toothache was soothed by an application of the powdered roots.

The husk, or cover, of plantain seeds contain psyllium gum, a mucilaginous substance that swells when added to water and serves as a natural bulk laxative, with the added benefit of reducing cholesterol levels, thereby lowering the risk of cardiovascular disease. The modern commercial laxatives, Metamucil and Perdiem, consist of the husk from plantain seeds.

Plantago species host caterpillars of the common buckeye butterfly (*Junonia coenia*).

Plantago lanceolata, narrow-leaved plantain.

POLYGONACEAE (Buckwheat Family)

Rumex hymenosepalus

CANAIGRE, DOCK, WILD RHUBARB, TANNER'S DOCK

Rumex has been a valuable source of food and medicine for centuries, with the genus name coming to us from the ancient Latin name for dock plants. The species name refers to the "membranous sepals" (*hymen*, "membrane," and *sepalum*, "sepal").

This three-foot-tall perennial inhabits sandy soil in disturbed areas of desert and grassland communities up to 6,000 feet elevation. The foot-long leaves rise from the short, ground-level stem and are waxy, wavy, and thick. Canaigre blooms from spring to early summer, with the flower head resembling that of grain sorghum as it matures. The flowers are green and have three "wings"; they mature into reddish-brown fruits (achenes).

The large leaves of canaigre, as well as the stems and seeds, are edible and have been used as food through many centuries. Though not cultivated in the true sense, the plant was allowed to grow among the crops and in the gardens of the ancestral peoples of New Mexico. It has a tart taste due to the presence of oxalic acid. The stems can be cooked in the same manner as rhubarb (hence one of its common names) and can also be boiled down to make jam. Several groups of Native Americans ground the seeds into a meal to make porridge or bread.

Rumex hymenosepalus, canaigre; field view of mature plant.

The roots are odorless and bitter, with the bitterness due to the presence of tannins, useful for tanning leather, as indicated by another of the plant's common names. Tannins, or polyphenolic compounds, also have medicinal value in healing damaged tissue. The Papago put pieces of the roots in their mouths to promote healing of sore gums. The Zuñi, Acoma, and Laguna ground the roots into a powder that they used on burns, rashes, and injuries to cure or prevent infections.

At one time, patent medicines made from *Rumex* roots were sold as antiseptics throughout the United States. Medical practitioners among the early settlers used it to treat skin diseases, syphilis, and anemia.

The pollen of this highly revered plant, honored among the Navajo as a Life Medicine, is sprinkled as a blessing on ceremonial objects. Master Navajo weavers boil the roots as a source of dye for their woolen yarn, achieving shades of brown, red, yellow, orange, and green.

The oxalic acid content of *Rumex* increases as the plant matures, and excessive consumption of the leaves by both humans and livestock can result in poisoning due to the deposition of calcium oxalate crystals in the kidneys and blood vessel walls. At a low level of intake, the microorganisms in the rumens of ruminants can adapt to the oxalates and use them for energy.

Canaigre is a larval host to several butterfly species, including lustrous copper (*Lycaena cupreus*), ruddy copper (*Lycaena rubidus*), and purplish copper (*Lycaena helloides*) butterflies.

Blue Flowers

ASTERACEAE (Sunflower Family)

Cichorium intybus

CHICORY

Cichorium is the Latin version of the Arabic name for "chicory," which was typically eaten during the month of January (in Arabic, *tybi,* hence the species name, *intybus*), when the Nile receded and crops began to grow. Chicory has a long history of use in the Middle East and Europe and was brought to America by early European settlers. It has become naturalized throughout the United States and is found in various localities in New Mexico, in disturbed areas such as roadsides and fields up to 7,000 feet elevation. It has only ray flowers, and the leaves are tiny and few on the five-foot-tall stems.

In the Mediterranean countries where chicory originated, the bitter leaves are used to complement certain foods and are eaten as greens after being cooked in several changes of water and being embellished with garlic and other bitterness-reducing agents. Some varieties are eaten fresh in salads, just like chicory's close relative endive. Ground chicory root is a time-honored substitute for, or adulterant of, coffee in places where coffee is prohibitively expensive.

Native Americans concocted a tea from the roots to use as a nerve tonic and made a root poultice to relieve the misery of cold sores. They also took advantage of a volatile oil concentrated in the roots that is effective in expelling intestinal worms and other parasites.

In Europe and the United States, chicory is grown commercially for the inulin stored in its roots. Inulin is a polysaccharide that is similar to starch but is made up of fructose rather than glucose. Because inulin's

Cichorium intybus, chicory; field view.

"starch" is not glucose based, it helps regulate blood sugar levels in diabetics. As a soluble dietary fiber, inulin increases laxation, reduces cholesterol levels, and improves overall intestinal health by providing a food source for beneficial bacteria in the intestines. As an ingredient in the food industry, inulin serves as a low-calorie substitute for sugar and fat in foods such as dairy products, meal-replacement beverages, baked goods, and meat products. It also helps prevent and inhibit colon and breast cancer through its effects on the immune system. One inconvenient side effect is increased flatulence, as the bacteria release gas while they are digesting the inulin.

BORAGINACEAE (Forget-me-not Family)

Mertensia franciscana

BLUEBELLS, FRANCISCAN BLUEBELLS, FRANCISCAN LUNGWORT

The genus name honors Franz Mertens, a German botanist and author, whose work bridged the 1700s and 1800s. The species name recognizes Arizona's San Francisco Peaks, which themselves were named for Saint Francis of Assisi.

These perennials, with stems as tall as three feet, are found in all montane zones of New Mexico, growing in rhizomatous colonies along streams in pine and aspen communities above 7,000 feet elevation. Blooming from May through August, the fragrant flowers change color with age, opening pink, progressing through lavender, and winding up blue. In the fall, bluebells' leaves join the parade of colors with subdued red and purple hues.

Bluebells are popular in woodland and mountain meadow landscapes. They are shade tolerant and need moist, acidic sandy or loamy soil.

CAMPANULACEAE (Bellflower Family)

Campanula rotundifolia

BLUEBELL, SCOTCH BLUEBELL, HAREBELL

Bluebell is native to America, but the same species is found in Scotland, where it is the symbol of the MacDonald clan. The plant at one time provided the blue dye used in the manufacture of tartans.

Campanula rotundifolia (little bell with round leaves) aptly describes the shape of the flowers and the basal leaves. However, the cauline (stem) leaves are long and narrow, like grass leaves.

Bluebell is delicate-looking, with slender stems about two feet tall holding several nodding, inch-wide flowers that bloom from June to September. This perennial grows in the western two-thirds of New Mexico, in moist soil in meadows or along roadsides in ponderosa pine to alpine zones.

During sweat lodge ceremonies, the Ramah Navajo used bluebell as a fumigant for various medical conditions, including eye problems and "deer infection," which causes a type of rheumatism. They also rubbed the plant on their bodies to protect themselves during hunts and to ward off witches.

Bees and hummingbirds pollinate bluebell, and it is also self-fertile. It spreads by rhizomes but also will grow easily from seed. Bluebell is a

Campanula parryi, Parry's harebell; note upturned flower.

charming addition to a native planting, especially in a rock garden, and will tolerate partial shade.

Also shown here is *C. parryi*, a close relative of *C. rotundifolia* with smaller, upright flowers that are borne individually on the stems, which have linear leaves.

FABACEAE (Bean Family)

Lupinus argenteus var. *heteranthus* (syn. *Lupinus caudatus*)

TAILCUP LUPINE

Scientific names often indicate interesting features of a plant. In this case, the quaint genus name prevaricates. The name *Lupinus* comes from the Latin word *lupus*, meaning "wolf," based on the mistaken belief that these plants deprive the soil of nutrients. In fact, the opposite is true; they fix atmospheric nitrogen through the assistance of *Rhizobium* bacteria housed with their root systems, thus enriching the soil.

The taxonomy of *Lupinus* is complex, due to much hybridization among members of the genus. The large and showy tailcup lupine is

Lupinus argenteus var. *heteranthus*, tailcup lupine; seedpods.

Lupinus argenteus, silver lupine; white color variant. Note gibbous calyx at flower base.

about as tall as it is wide. Its less robust, but equally beautiful cousin, silver lupine, also shown here, has a gibbous calyx, in which part of the sepal cup bulges back toward the stem. Both inhabit the northern and western halves of New Mexico from 7,000 to 10,000 feet elevation. At scarcely five inches tall, shortstem lupine (see photo) is a scaled-down version of its larger cousins, inhabiting sandy soil in pinyon-juniper woodlands.

The Navajo used lupine as a nausea-inducing substance to clear the body in preparation for ceremonies. They treated bruises and boils with a

Lupinus brevicaulis, shortstem lupine.

poultice of ground roots. The Ramah Navajo soothed the intensely itchy rash of poison ivy by applying crushed leaves or a wash made from a leaf infusion. One traditional use of lupine was as a remedy for infertility, particularly if a female child was desired.

Although lupines are gorgeous, they are also deadly, with high concentrations of the toxic quinolizidine alkaloids, especially anagyrine. These toxins cause great economic losses from the death of sheep and birth defects in cattle, particularly crooked calf disease, in which the joints are immobile or crooked. The toxins begin to accumulate in the plants during flowering in June through October, with the highest concentrations stored in the seeds, making this stage the most dangerous for livestock. The alkaloid content drops as soon as the pods shatter and drop the seeds.

Elk, deer, antelope, various bird species, and small mammals consume various parts of the plant, and smaller animals use it for cover. They do not seem to suffer the ravages of the toxins, or they may know to avoid the plant when the toxins are at their highest levels. *Lupinus argenteus* is a caterpillar host for several butterfly species, including Queen Alexandra's sulphur butterfly (*Colias alexandra*), and is a nectar source for dreamy duskywing (*Erynnis icelus*), arrowhead blue (*Glaucopsyche piasus*), and Boisduval's blue (*Plebejus icarioides*) butterflies.

IRIDACEAE (Iris Family)

Iris missouriensis

ROCKY MOUNTAIN IRIS, WESTERN BLUE FLAG, MISSOURI IRIS

We are so accustomed to the iris as a garden flower that we may forget there are wild species and that it was named for Iris, the Greek goddess of the rainbow. The species name, *missouriensis*, is based on an early collection close to the Missouri River's headwaters in 1834.

Masses of stunningly gorgeous blue to purple flowers bloom from May to July in the high mountain meadows from 6,000 to 10,500 feet elevation. They are found in the western and northern montane regions of New Mexico and in the central region as far south as the Sacramento Mountains. The plants are about two feet tall and spread rapidly by underground rhizomes in moist soil or shallow bodies of water, forming huge colonies. Lacking an aboveground stem, the leaves rise directly from the rhizomes.

Rocky Mountain iris has powerful toxic agents in its rhizomes that cause violent vomiting and diarrhea. The leaves have a similar though

milder effect. Both livestock and wildlife avoid it, so populations increase in overgrazed areas.

The toxin is the bitter-tasting iridin, a glycoside compound that is a severe irritant to the gastrointestinal tract and skin. Iridin also has pharmacological properties, including anti-inflammatory, antitumor, and antiviral activity. In spite of the bitterness and bad smell, the rhizomes served as a source of medicine for Native Americans. They were mashed and placed in a tooth cavity or beside an aching tooth to relieve pain, or on aching joints to relieve rheumatic discomfort. The rhizomes were brewed into a bitter-tasting tea to treat venereal diseases, and a salve made from

Iris missouriensis, Rocky Mountain iris; field view.

Iris missouriensis, Rocky Mountain iris; three-chambered seedpods.

the rhizomes was rubbed on sores caused by syphilis. The dilute tea was also drunk for urinary tract infections and stomachache, and it was used as an eyewash and as eardrops to treat earache. The Shoshone and Paiute made a paste of iris seeds to heal burns. The Zuñi poulticed newborns with chewed roots to increase their strength.

Rocky Mountain iris attracts hummingbirds and an assortment of bees.

IRIDACEAE (Iris Family)

Sisyrinchium desmissum

STIFF BLUE-EYED GRASS

Sisyrinchium is a mystery name, and examination of its origin gives us two interesting choices. On the one hand, we find that *sisyr* is the Greek word for "hairy goatskin," perhaps referring to the webbed covering of the roots. On the other hand, we find a widely purported derivation from *sus*, Latin for "pig," and *rhynchos*, or "nose," referring to pigs using their noses to unearth the roots of this plant. Given the roots' extreme toxicity, the second possibility seems less valid.

Blue-eyed grass is found in central and western New Mexico, in moist meadows and temporary ponds in desert to spruce-fir communities from 2,800 to 9,500 feet elevation. It blooms from April through October. The leaves resemble grass (hence the common name) and rise directly from the rootstock.

Blue-eyed grass has toxic properties similar to those of *Iris missouriensis* (see the previous entry), with a glycoside concentrated in the roots that causes violent vomiting and diarrhea in livestock. Fortunately, the animal must consume large quantities of the plant before poisoning occurs, reducing the likelihood of harm.

LINACEAE (Flax Family)

Linum lewisii var. *lewisii*

LEWIS FLAX, PRAIRIE FLAX, BLUE FLAX

Fiber from flax, known as linen (from the Greek *linon*), has been an important source of cordage and textiles since ancient times in the Middle East and North Africa. The species name was given in honor of Meriwether Lewis, of the famous Lewis and Clark Expedition to the Pacific Northwest in 1804–06.

Linum lewisii var. *lewisii*, blue flax; field view.

This perennial blooms from April through September throughout New Mexico in pinyon-juniper to Douglas fir communities up to 9,000 feet elevation. A number of insects visit flax early in the day, but the flowers close for business by noon.

Linum lewisii is native to North America and is similar to the cultivated species *L. usitatissimum*, from which we get not only fiber from the stems but also oil from the seeds. Flaxseed oil, also known as linseed oil, is high in alpha-linolenic acid, an essential fatty acid. In addition to its use in dietary supplements, linseed oil is used in paint, ink, and varnish. Crushed flaxseed is also marketed as a dietary bulking agent for constipation, based on the water-absorbing properties of its mucilage content.

Native Americans used the pleasant-tasting, nutritious seeds to add oil and protein to their diet. Even more important than the food value were the strong, flexible fibers of the stems, which were used to make string, ropes, and fishing lines.

Linum lewisii var. *lewisii*, blue flax; seedpods.

Native Americans boiled the roots to obtain a shampoo. They also made an infusion of the flowers, leaves, and stems and used it to wash their skin. This infusion was particularly popular among young girls, who thought it gave them a pretty complexion. Older people concerned about losing their hair shampooed with the whole plant.

The Navajo applied warm, cooked leaves to goiters and other swollen areas, and they used the plant to treat burns.

Livestock and big game animals feed on the foliage. Birds, especially grouse, consume the seeds. Lewis flax is drought hardy and tolerates some shade, making it useful as well as ornamental in natural landscape plantings. It reseeds very easily, with seed establishment enhanced by the sticky mucilage, which anchors the seeds in the soil and retains water during the germination process. The seed is commercially available.

Lewis flax is used in the reclamation of degraded land, such as mining sites and roadsides, and for rehabilitation of rangeland. It is very competitive with noxious weeds, such as knapweed (*Rhaponticum repens,* see p. 185) and cheatgrass (*Bromus tectorum*), effectively reducing their populations.

For information about other flax species, see p. 32.

PLANTAGINACEAE (Plantain Family)

(formerly in Scrophulariaceae)

Penstemon strictus

ROCKY MOUNTAIN BEARDTONGUE, ROCKY MOUNTAIN PENSTEMON

The name *Penstemon* tells us that the members of this genus have "five stamens." The species name, *strictus,* Latin for "upright," describes the straight stems, which are well packed with blue-violet flowers.

Rocky Mountain beardtongue is found in northern and western New

Penstemon strictus, Rocky Mountain beardtongue; field view.

Mexico in pinyon-juniper to spruce-aspen zones from 6,000 to 11,000 feet elevation. This three-foot-tall perennial grows in sandy or rocky soil and blooms in June and July—much too short a season for such stunning flowers!

Rocky Mountain beardtongue is an essential member of natural landscapes or gardens with well-drained soil. It will tolerate some shade. This and other penstemons are useful in wildflower mixes for land reclamation and erosion control. Seed is commercially available; root division is another means of propagation.

Wildlife such as antelope and deer use penstemons as forage, and birds consume the seeds. Penstemons are melittophilous, a lovely word meaning that bees are the main pollinators. Butterflies, various flies, and even hummingbirds also visit *Penstemon strictus*.

For information about other *Penstemon* species, see pp. 20 and 248.

Penstemon strictus, Rocky Mountain beardtongue; with tiger swallowtail (*Papilio glaucus*).

POLEMONIACEAE (Phlox Family)

Ipomopsis longiflora

PALE TRUMPETS, LONG-FLOWERED GILIA, BLUE STARFLOWER

The genus name, *Ipomopsis,* was created in 1803 by André Michaux, a French botanist and explorer who recognized that the flowers were similar to those of *Ipomoea,* a genus in the morning-glory family. The species name, *longiflora* (long flower), refers to the length of the floral tube. Blowing its trumpets to announce the arrival of spring, this delicately beautiful member of the phlox family came by its common name "pale trumpets" very naturally.

The spindly, fragile plant has many branches, each tipped with a pale blue flower. It grows throughout much of New Mexico in sandy or rocky soil in the pinyon-juniper zone below 7,000 feet elevation.

The Tewa mixed the flowers with water to produce a sudsy cleanser for wounds, boils, and other sores, while the Ramah Navajo used the solution as a ceremonial eyewash. The Navajo swabbed newly castrated colts with the cleanser to promote healing. *Ipomopsis longiflora* also has internal cleansing capabilities and was used as a purgative for both medicinal and ceremonial purposes.

Pale trumpets is a Life Medicine of the Navajo and is favored for treating stomach problems in both humans and livestock. People chew the plant with salt to relieve heartburn. In the past, the Ramah Navajo used the plant as a hair tonic to prevent baldness, while the Zuñi used a flower poultice to remove the hair of newborns. Navajo women valued pale trumpets during childbirth as an aid in delivering the placenta and as a personal wash afterward to prevent postpartum infection.

Ipomopsis pumila, dwarf ipomopsis; whole plant.

The plant is more robust than it appears and is a good candidate for a natural or xeric landscape. Pale trumpets has a prolonged flowering period, from March to October. The two-inch-long flowers have deep, narrow floral tubes that fit the mouthparts of the moths that serve as their pollinators, of which the most important is the white-lined sphinx moth (*Celerio lineata*). The plant's late-hour blooming schedule is timed to match the nocturnal habits of moths, which are attracted to the pale blue blooms that "glow" at dusk or in the moonlight. Pale trumpets grows readily from seed, making it easy for the home xeriscaper to bring this ecological partnership into a domesticated framework.

Ipomopsis longiflora has a dainty little cousin, *I. pumila* (see photo), that grows in similar plant communities at the same elevations in central and western New Mexico.

For information about another *Ipomopsis* species, see p. 21.

Purple Flowers

AMARYLLIDACEAE (Amaryllis Family)

(formerly in Liliaceae)

Allium acuminatum

TAPERTIP ONION

Wild onions fall under the same nomenclatural umbrella as their more pungent cousin, garlic, whose Latin genus name is also *Allium*. The species name of tapertip onion, *acuminatum*, describes the long, "tapering points" of the sepals and petals, which look like identical twins.

These dainty wildflowers look lovely even when they cover your lawn and are scurrilously called "weeds." They bloom from March to June, growing in calcareous soil in grasslands up to about 5,000 feet elevation.

The long, linear leaves of tapertip onion and its close cousin *Allium cernuum*, "nodding onion" (see photos), rise from an underground bulb, which is like that of their commercial cousins, though on a much smaller scale. This bulb is covered with a fibrous "fabric" that has been excavated at many ancient American Indian archaeological sites, indicating that wild onion's use in the Southwest was extensive. Nodding onion shares many characteristics and properties with tapertip onion but grows at higher elevations, up to 9,000 feet.

Wild onions, along with plants in the asparagus family, *Agave* (p. 8) and *Dasylirion* (p. 270), and others in the sunflower family, contain sig-

Allium cernuum, nodding onion; whole plant.

Allium cernuum, nodding onion; immature inflorescence wrapped in spathe.

nificant amounts of inulin, a type of starch consisting of fructose units, which helps control diabetes and cholesterol levels.

Onions, whether wild or tame, contain various sulfur compounds that impart a characteristic onion scent and also deliver medicinal effects. Among Native Americans, onion juice rubbed on the skin served as an insect repellent. After the fact, crushed wild onion bulbs were placed on bee or wasp stings to relieve the pain. Crushed bulbs were also used to poultice carbuncles, the old-fashioned name for a skin infection caused by *Staphylococcus aureus*. The Blackfoot made a tea from wild onions to treat coughs and to relieve nausea and vomiting. For colds and sinus headaches, they charred the bulbs and inhaled the smoke. They also used wild onions to treat ear infections and intestinal worms. Early pioneers made wild onion poultices to treat pneumonia and chest colds.

One of the sulfur compounds in wild onions, N-propyl disulfide, is not so benign. Cats, dogs, and horses can be poisoned by overconsumption of wild onions due to the effects of this compound, which destroys red blood cells. If you find your pet with "onion breath," head for the vet.

Native Americans extracted a brownish-orange dye from wild onions to color yarn for weaving.

Wild onions can be cultivated from both seed and bulbs and add spring color to a xeriscape garden. They are not popular in deer diets and can be used to keep deer away from gardens.

ASTERACEAE (Sunflower Family)

Arctium minus

COMMON BURDOCK, BURWEED, BUTTONBUR

Burdock fruiting structures are Velcro in action, waiting for an opportunity to hitch a ride. The genus name means "like a bear," in reference to the bear claw-like, hook-tipped phyllaries that are located under the many disk flowers.

This six-foot-tall, late-blooming biennial flowers from August to October and has exceptionally large lower leaves that can reach almost a foot in length and width. It is found in northern New Mexico in disturbed areas along roadsides and waste places up to about 6,500 feet elevation. The bur can be confused with those of *Xanthium*, or cocklebur (see p. 144), but the leaves make them easy to distinguish. *Arctium* leaves are large and soft, while *Xanthium* leaves are smaller and either coarse textured and deltoid (*Xanthium strumarium*) or shiny and linear (*X. spinosum*).

Common burdock was introduced from Europe and Asia and has now become widespread throughout the United States. Like *Xanthium*, burdock is classified as a noxious weed in some states.

Its soft young leaves are eaten as a cooked green vegetable. The stems are peeled and eaten either raw or cooked. The roots are also edible, cooked in a soup or dried for future use. They contain a significant amount of inulin, which acts as a dietary fiber and helps regulate blood sugar levels in diabetics.

In the medicine chest, the leaves have multiple uses. They were used in a tea to treat whooping cough, relieve bee stings, and cure earache and as a poultice to treat headache and skin sores. Burdock also acts as a diuretic and inhibits kidney stone formation. The roots can be used to heal skin sores and pimples, treat stomach pain and venereal diseases, and ease labor pains.

Burdock seeds and leaves contain arctigenin, a lignan compound that shows activity against HIV and other viruses, as well as against tumors. Burdock also demonstrates activity against gram-negative bacteria.

Common burdock is a larval host for the American lady butterfly (*Vanessa virginiensis*).

ASTERACEAE (Sunflower Family)

Cirsium neomexicanum

NEW MEXICO THISTLE, LAVENDER THISTLE

The genus name, *Cirsium*, is derived from the Greek word for "thistle," *kirsion*, which comes from the word *kirsos*, meaning "a swollen vein," for which thistles were used as a remedy. Asa Gray named this species in 1853 for the territory in which it was collected—our own New Mexico.

There are several thistles found in New Mexico, some introduced and others, such as New Mexico thistle, native. Unlike most other thistles, this species is not weedy. But like all thistles, its flower heads have only disk flowers.

New Mexico thistle is found in the western half of New Mexico, growing in sandy or other well-drained soil in disturbed areas and in

Cirsium neomexicanum, New Mexico thistle; close-up of immature flower head. Note cobweb-like white hairs.

grassland and pinyon-juniper zones up to 7,000 feet elevation. The plant can reach as tall as six feet, with the stalks from the previous year hanging around, not yet willing to make the transition to compost. The lavender, pink, or sometimes white flowers appear in April and May from the midst of very spiny phyllaries that have cobweb-like hair draped among them like Halloween decorations (see photo).

The Paiute and other Indian tribes ate the stems of thistle. It is a Navajo Life Medicine. A root tea was used as an eyewash for both people and livestock, and to treat fever and headache. The Zuñi made a contraceptive from crushed roots. The people of the San Juan Pueblo made a poultice of thistle seeds to cure boils.

Thistle seeds are popular with birds, especially goldfinches, which also use the fluffy bristles on the seeds to make their nests more comfortable. The flowers are a nectar and pollen source for bees, hummingbirds, and butterflies, including black swallowtail (*Papilio polyxenes*), common wood nymph (*Cercyonis pegala*), painted lady (*Vanessa cardui*), and pipevine swallowtail (*Battus philenor*) butterflies. *Cirsium neomexicanum* is also host to the caterpillars of mylitta crescent (*Phyciodes mylitta*) and painted lady butterflies.

Lavender thistle must contend with other insects that are undeterred by its spiniferous character and use it as a food source by consuming it from within. These include the artichoke plume moth (*Platyptilia carduidactyla*), which destroys the stems and crown region at ground level.

A native cousin, *Cirsium ochrocentrum* (see photo), is a perennial found throughout New Mexico. The yellow-spine-tipped leaves are covered with matted white hairs. To treat diabetes, the Zuñi drank a tea made

Cirsium ochrocentrum, yellowspine thistle; flowering head.

Cirsium ochrocentrum, yellowspine thistle; gone to seed.

from the roots three times a day. They also drank a tea made from the whole plant to treat syphilis. The Zuñi were egalitarian in their contraceptive practices: both partners drank a tea made from the roots to space the arrival of their children.

A European cousin of New Mexico thistle, *C. vulgare* (see photo), arrived in the United States in the late 1700s and has become a noxious weed in most regions of New Mexico, particularly in the northwest counties. Its wicked spininess is slightly redeemed by the story of its key role more than seven hundred years ago, in 1263, in the deliverance of Scotland from the invading Vikings. Knowing they were under threat of attack, the Scots gathered a multitude of thistles and piled them on the beach where the Vikings were likely to arrive. When the Vikings landed under cover of darkness, they took off their shoes, thinking to sneak up quietly on the Scots. But when their feet encountered the thistles, they announced their presence with shouts of pain and anger. The Scots welcomed them with a great show of force and sent them packing. Now the Battle of Largs and the thistle of salvation are immortalized by Scotsmen in songs, and the thistle is honored as the national flower, in the motto "Who Dares to Meddle with Me," and as a chivalric order, the Order of the Thistle.

Cirsium vulgare, bull thistle; flowering head.

ASTERACEAE (Sunflower Family)

Dieteria canescens (syn. *Machaeranthera canescens*)

HOARY-ASTER, PURPLE TANSYASTER, SAND DAISY

This handsome species has been caught in the crossfire of nomenclatural changes. At the moment, the genus name has reverted to a much earlier one, *Dieteria*, given by the well-known botanist Thomas Nuttall in the early 1800s. The name was apparently intended by Nuttall to refer to its biennial growth habit (*di,* "two," and *etesios,* "yearly"). The species name, *canescens* (turning to gray), refers to the short white hairs on the leaves and stems.

Purple tansyaster blooms in late summer and fall, growing in a variety of soil types in communities of saltbush, sagebrush, and aspen, making showy displays along roadsides and in disturbed areas. The flower heads have few or no sticky glands and therefore look clean and tidy. Only the recurved tip of the phyllary is dark green, with the remaining three-quarters tan or light green. The flowers are light purple or lavender-pink. As they age, the ray petals curl backward, looking like a head full of blue curlers.

Several New Mexico tribes used tansyaster to treat stomach ailments. The Zuñi prepared a tea for an emetic to alleviate upset stomach, while the Pueblo people along the Rio Grande chewed the fresh flowers for the same purpose. The Ramah Navajo dried and crushed the plant to use as a snuff to clear the nose and throat. This powder was also dusted over a freshly washed face to absorb skin oil and prevent pimples.

Dieteria canescens, sand daisy; white version at White Sands National Monument.

Tansyaster is a pioneer species on rangeland and disturbed sites and is able to compete with invasive weeds such as knapweed and cheatgrass. Seed is commercially available.

Purple tansyaster is a nectar source and caterpillar food for sagebrush checkerspot (*Chlosyne acastus*) and field crescent (*Phyciodes pulchella*) butterflies, and a nectar source for cabbage white (*Pieris rapae*) and margined white (*Pieris marginalis*). Other insect visitors include several species of sweat bees, bee flies, and honeybees. These insect associations with tansyaster are an important food source for sage grouse during rearing of their young.

A similar-looking cousin, *Dieteria bigelovii*, has very sticky glandular hairs on the phyllaries and peduncles (the "stems" that hold the flower heads). This "glue" remains on your hands or clothes after touching them. Indeed, this stickiness catches small sand particles and makes the plant look dirty and unkempt. *Dieteria bigelovii* plants are usually slightly taller and more robust-looking than those of *D. canescens*. They also have fewer flowers, which are a deeper purple or bluish color. That being said, the two species hybridize, and in regions where their ranges overlap, it is difficult to clearly differentiate them.

ASTERACEAE (Sunflower Family)

Liatris punctata

DOTTED GAYFEATHER, PRAIRIE SNAKEROOT, BUTTON SNAKEROOT

The leaves of *Liatris punctata* are "dotted" (*punctata*) with oil-bearing glands. The derivation of two of its common names references the belief that it provides protection from snakes.

Gayfeather is a member of mature native prairie communities as well as of pinyon-oak woodlands. It grows slowly, remaining as a rosette for at least a year while its deep taproot descends to about three feet. It prefers well-drained calcareous soil, surviving for up to five years, but succumbs quickly in heavy soils. Gayfeather can reach up to three feet tall and flowers in late summer through fall.

Gayfeather's corms (small, fiber-covered bulblets) are a means of propagation and also serve as a source of food and medicine. The corms contain inulin, a fructose polysaccharide that is digested slowly and is useful in helping control blood sugar levels in people with diabetes. Some say the pounded corms taste like carrots; others think they have a pine-like, turpentine scent and taste. The discrepancy is probably due to the time of year and amount of rainfall the plants received.

This plant contains sesquiterpene lactones and pyrrolizidine alkaloids, some of which have antileukemia activity and anti-inflammatory properties. In the past, the corms were used as a sore throat treatment, a diuretic, and a gonorrhea medicine. The corms were also roasted and the smoke inhaled as a treatment for headache or nosebleed. The entire plant was powdered to make a tea for heart ailments.

Liatris punctata, dotted gayfeather; with close-up of flowers and bombyliid bee fly.

The leaves have a faint scent of vanilla and were used to protect clothes from moths and to flavor tobacco.

In a natural or waterwise landscape, gayfeather provides strikingly attractive fall color. It is also tolerant of shade. It is now popular as both a cut flower and a dried flower in the floral industry.

Liatris punctata is an important host to the hemiparasite *Castilleja integra* (see p. 18), which selectively takes up pyrrolizidine alkaloids from the *L. punctata* root system, thereby enhancing its own defense system against predators.

Gayfeather has little forage value, but domestic livestock and wildlife may browse it when the plants are tender. Although the plant contains alkaloids, the concentration is so low as to pose little risk of poisoning.

Liatris punctata is an important nectar source for the painted lady butterfly (*Vanessa cardui*), as shown in the main photo, and the Leonard's skipper butterfly (*Hesperia leonardus*), as well as for a vast host of other butterflies and insects, such as the bombyliid bee fly (see photo).

ASTERACEAE (Sunflower Family)

Machaeranthera tanacetifolia

TAHOKA DAISY, TANSYLEAF ASTER, TANSYASTER

Tahoka daisy's ponderous scientific name tells us that it has "sword-shaped" appendages on the anthers (*machaeranthera*) and that it has finely divided, "tansylike leaves" (*tanacetifolia*). While its common name commemorates the small West Texas town of Tahoka, where the seeds were first gathered for commercial production, it blankets New Mexico with a stunning purple carpet just as beautifully as it does Texas.

Tahoka daisy is found in disturbed areas in grassland, chaparral, and pinyon-juniper communities up to 5,500 feet elevation. This fast-growing annual continues to be popular among home gardeners and waterwise landscapers for its extreme drought tolerance, willingness to grow and flower in very poor soil, and unending production of flowers from early spring to just past frost. Tahoka daisy has a multitasking skill that few plants have: it can simultaneously produce fresh flowers and mature seeds throughout the growing season. Of course, it develops a rather frumpy look with all the fuzzy seeds hanging about the plant, but that is easily taken care of with a quick buzz cut with the clippers, to begin the cycle anew.

Machaeranthera tanacetifolia, tansyaster; mature seed head. Some "children" have already left home.

The Navajo made a tea from the entire plant to treat stomachache, and they made a snuff from the dried roots to induce a sneeze that would clear a congested nose. The Pueblo Indians chewed the fresh flowers to relieve stomachache. The Hopi used *Machaeranthera* species as a stimulant for women suffering a difficult childbirth.

ASTERACEAE (Sunflower Family)

Rhaponticum repens (syns. *Centaurea repens* and *Acroptilon repens*)

RUSSIAN KNAPWEED, HARDHEADS, TURKESTAN THISTLE

Knapweed's pretty purple flowers belie its wicked weedy ways. In the early 1900s, it entered the United States in alfalfa seed arriving from the Caucasus region, where it has long been regarded as a perniciously destructive weed in farmland. It has spread so rapidly that several states, including New Mexico, classify it as a Class B noxious weed, meaning that it is a nonnative species that is abundant in some areas and at risk of spreading to others and needs hypervigilance to control its spread into uncontaminated areas.

The genus name, *Rhaponticum*, points to the region in Russia where the plant originated: the Volga River, Black Sea, and Caspian steppes. The species name, *repens*, indicates that this weed has stems that creep and root at the nodes, giving it exceptional skill in forming dense colonies. Its takeover strategy includes an arsenal of allelopathic compounds that restrict the growth of other plant species near it.

In addition to overwhelming crops in agricultural land, it also contains sesquiterpene lactones that are very toxic to horses, causing irreversible brain damage. Livestock avoid it because of its smell and bitter taste, but if suitable grazing is not available, they will resort to eating it.

Knapweed proliferates in any disturbed area, from roadsides to farmland to overgrazed rangeland. Plants in bloom look quite lovely with their dainty flowers, which appear in May through September. Insects are not put off by knapweed's scurrilous character and partake readily of its nectar and pollen. Large wildlife, such as deer and bighorn sheep, browse Russian knapweed, and birds and small mammals eat the seeds.

ASTERACEAE (Sunflower Family)

Vernonia marginata

PLAINS IRONWEED

The genus name honors the English botanist William Vernon, who made collections in eastern North America in the late 1600s. *Marginata* refers to the white, eyelash-like fringe along the margins of the purple phyllaries surrounding the flower head. The stems are tough, as the name "ironweed" indicates.

Vernonia's violet flowers are a pleasant visual respite from the ubiquitous "yellowscape" of the summer months. They bloom from July into the fall, growing in sandy soil in disturbed areas in pinyon-juniper communities up to about 6,000 feet elevation, mainly in eastern New Mexico. The plants are often found in large colonies because they reproduce by rhizomes.

Ironweed is useful in waterwise landscapes not only for its drought tolerance but also because it hosts so many insects, in particular larvae of the American lady butterfly (*Vanessa virginiensis*).

Where ironweed grows in pastures, its unpalatable chemical content protects it from grazing animals. Significant populations of *Vernonia* in rangeland, therefore, indicate poor management. The seeds contain significant quantities of vernolic acid, an important oil with a potential market in the manufacture of paint and pharmaceuticals and in other industrial applications.

BORAGINACEAE (Forget-me-not Family)

Cynoglossum officinale

HOUND'S TONGUE, BEGGARS-LICE, DOGBUR

Linnaeus observed that the leaves of this plant were shaped like a "dog's tongue," but he made it sound more sophisticated by calling it *Cynoglossum,* which means the same thing. It was widely used as "official" medicine, as the species name indicates, in its home countries in Europe.

Hound's tongue is believed to have arrived in Canada as a stowaway in crop seed in the 1850s. Within 130 years, it had spread across most of

Cynoglossum officinale, hound's tongue; field view.

North America and, thanks to its toxicity and aggressive propagation by seed, is now classified as a noxious weed in several states.

This biennial with a stout black taproot forms a rosette with soft, hairy leaves in the first year. In the second year, the flower stalk can reach four feet tall, producing reddish-purple flowers from June through August. Each flower in turn produces four heavily barbed nutlets, as shown in the photo, that latch onto animals or clothes. It is a plant of waste places and roadsides, found mainly in the northern third of New Mexico from 5,000 to 8,000 feet elevation. Bumblebees provide pollination services, and butterflies, bees, and other insects dine on its nectar and pollen.

The bad odor and taste imparted by the toxic compounds cynoglossine and consolidin render hound's tongue unsuitable for food, except as a red colorant in wines and some food products.

The plant has medicinal properties within its chemical profile. The outer layers of the roots contain fat-soluble red pigments that inhibit the growth of bacteria and tumors and promote wound healing, making this species useful in home remedies to treat burns, hemorrhoids, skin problems, and fever.

Cynoglossum officinale, hound's tongue; note barbs on nutlets.

The odor of hound's tongue is repulsive to most animals. Even so, mule deer, domestic sheep, rabbits, and pheasants make it a minor source of food, and do so with evident impunity. It is highly toxic to cattle and horses through the action of several pyrrolizidine alkaloids, which are at their highest concentration in the rosette stage. These toxins damage the liver, and poisoning can be chronic or acute, depending on how much is consumed over a given period of time.

BORAGINACEAE (Forget-me-not Family)

(formerly in Hydrophyllaceae)

Nama hispidum

SANDBELLS, PURPLE NAMA, BRISTLY NAMA

This petite beauty is easily overlooked because of its short stature, usually eight inches or less. It inhabits all regions of New Mexico, except the northeast, growing in sandy, gravelly, or caliche soils in desert scrub below 7,000 feet elevation. The plants are very sociable and grow in dense swaths, with the flowers putting on their show, both visual and olfactory, from March through November.

The genus name is Greek for "a spring" or "a stream," possibly referring to its propensity for growing along washes in the desert, or maybe because a big area covered by sandbells looks like a desert watering hole. The species name tells us that the plant is covered with stiff hairs, hence the common name "bristly nama."

With its ability to withstand arid conditions, sandbells makes an attractive border plant in a rock garden.

The foliage is browsed by cattle and white-tailed deer.

BORAGINACEAE (Forget-me-not Family)

(formerly in Hydrophyllaceae)

Phacelia integrifolia

GYPSUM PHACELIA

Like other flowers in the forget-me-not family, gypsum phacelia's flowers are arranged in a coiled cyme, resembling a scorpion's tail. This "bundle" (*phacelus*) of flowers gives the plant its genus name. The margins of the leaves are scalloped, but since they do not have deep divisions, they are referred to as *integrifolia* (whole leaves). The leaves and stems have sticky glands that share an unpleasant smell with the world.

Phacelia is easily recognized by the long stamens extending out from the flowers and petal colors that are typically bluish, purplish, or pinkish. The lacy-looking flowers bloom from March through May. *Phacelia* grows in sandy or loamy soil in desert and grassland communities. The plants are drought tolerant and very suitable for natural landscaping projects.

Native Americans ate *Phacelia* leaves as a cooked green vegetable, either boiled or fried. The Acoma and Laguna made an infusion to drink

Phacelia integrifolia, gypsum phacelia; whole plant.

for sore throat or to rub on swellings and sprains. The Zuñi mixed the powdered roots in water for treating irritated skin and rashes.

Phacelia attracts butterflies and is palatable to cattle and white-tailed deer.

DIPSACACEAE (Teasel Family)

Dipsacus fullonum

FULLER'S TEASEL

Dipsacus fullonum's connate leaves join the stems to form a cup large enough to hold water and invite you to quench your "thirst" (*dipsa*) with elixir of *Dipsacus*! Some Native American tribes used this liquid to soothe irritated eyes.

Fuller's teasel was important for the woolen textile industry until the industrial revolution in the 1700s began to reduce the demand for its services. The flower head's stiff bristles are strong enough to raise the nap

Dipsacus fullonum, fuller's teasel; field view.

on woolen cloth and to straighten it. *Fullonum*, the species name, refers to fullers, people who fulled cloth, which is the laborious process of thickening the fabric with moisture, heat, and pressure.

This species was introduced from Europe and was cultivated from about 1840 in New York, subsequently spreading across the United States via the extensive road network. Today it is seen most commonly along highways. The ten-foot-tall plants bloom from July through September and attract a wide range of insects. Birds and small mammals enjoy the largesse of teasel's prolific seed production.

The roots are used to treat warts, ulcers, wounds, and jaundice. The Iroquois made an infusion of teasel leaves for acne. The stiff inflorescence is attractive in dried flower arrangements.

FABACEAE (Bean Family)

Astragalus mollissimus

WOOLLY LOCOWEED

Among the tarsal bones in the foot is the astragalus bone, which is kidney-shaped like the seeds of locoweed, which derives its scientific name from this bone. Woolly locoweed has extremely soft, hairy leaves, as the species name, *mollissimus* (soft to a superlative extent), reveals.

Woolly locoweed is one of the most widespread *Astragalus* species in New Mexico, growing in open, rocky areas in every plant community up to about 8,500 feet elevation. The densely pubescent leaves are

Astragalus mollisimus, woolly locoweed; whole plant.

Astragalus nuttallianus, turkey pea.

an identifying characteristic, as are the short, plump, hairy seedpods, with their cute little curly pigtail. The plant flowers from April through midsummer.

As the most prevalent *Astragalus*, woolly locoweed has become the species most associated with locoism, or crazy behavior that follows its consumption. The craziness results from the toxin swainsonine, which dramatically affects the brains and nervous systems of both ruminants and horses. Horses are more sensitive to its effects, and their resulting extremely erratic behavior makes them too dangerous to ride.

Woolly locoweed is host to the caterpillars of Queen Alexandra's sulphur (*Colias alexandra*) and arrowhead blue (*Glaucopsyche piasus*) butterflies.

Not all *Astragalus* species are toxic, although about 80 percent of them are. Of the few "safe" ones, *Astragalus nuttallianus* (see photo) is quite common and is useful as forage. To be on the safe side, however, *do not* act loco and experiment with any plant even remotely resembling *Astragalus*.

For information about other *Astragalus* species, as well as the closely related genus *Oxytropis*, see pp. 305 and 311.

FABACEAE (Bean Family)

Dalea purpurea

PURPLE PRAIRIE CLOVER

The genus name, *Dalea*, honors Samuel Dale, a British botanist. The species name, *purpurea*, tells us that the flowers are "purple," and the common name confirms that.

Purple prairie clover is a subshrub, with many long flowering stems rising from the crown and daintily festooned with pinnately compound

Dalea purpurea, purple prairie clover; as seen from above.

leaves. The flowers bloom from June through September, with each inflorescence resembling a long, purple, frilly thumb. The seedpods are tiny and contain a single seed.

Purple prairie clover is an important part of the Great Plains grasslands and is found throughout eastern New Mexico and to some extent in the west. Ecologically, it is considered a mid- to late-successional species, part of the climax grassland community, but it can also be a pioneer species on roadsides and in other disturbed areas.

The plant's roots have a pleasantly sweet flavor and were chewed like gum by several Indian tribes, including the Lakota and Comanche. The leaves can be steeped to make a refreshing beverage.

Studies have confirmed the presence of antibacterial and fungicidal compounds in *Dalea purpurea*, which explains the medicinal value Native Americans placed on it. Montana Indians steeped crushed leaves in hot water and then applied them to fresh wounds. The Navajo used the plant to treat pneumonia.

Purple prairie clover is high in protein and very palatable to livestock and mammalian wildlife and will therefore decrease in an overgrazed pasture. It also fixes nitrogen in the soil.

The whimsically decorative effect of this plant makes it of great interest in a native landscape. All manner of insects find the flowers irresistible, from bees to walking sticks (see main photo). Purple prairie clover is host to larvae of the southern dogface butterfly (*Zerene cesonia*). Several USDA programs in different states have released locally adapted varieties for use in roadside plantings and as part of seed mixtures to help restore native vegetation.

For information about other *Dalea* species, see pp. 242 and 308.

FABACEAE (Bean Family)

Sophora secundiflora

MOUNTAIN LAUREL, MESCAL-BEAN

A mountain laurel in luxuriant, fragrant, purple bloom is one of those "we must get a closer look" trees. The red seeds are equally attention grabbing.

The genus name given by Linnaeus is derived from *safraa*, Arabic for "yellow," referring to yellow-flowered members of this genus. In his typical style, he apparently also intended a play on words with its similarity to the Greek word for wisdom, *sophia*. Indeed, some Native Americans sought a form of spiritual wisdom from this species through the Red Bean Dance. *Secundiflora* means that the flowers are arranged on one side of the flower stalk. The "mescal-bean" appellation probably came about because of similar out-of-one's-mind effects from *Sophora*'s red beans and the *Agave* drink.

Mountain laurel is a slow-growing evergreen tree that reaches about fifteen feet tall. It inhabits the southeastern New Mexico mountains, growing in limestone soil up to 5,000 feet elevation. When in bloom during March and April, the purple flower clusters share their potent scent of grape soda pop far and wide. The young seedpods, white and hairy and about the size of a pinkie, look like piglets nestled up for nursing. The mature fruits contain bright red seeds that are poisonous but make beautiful necklaces.

Deadly toxic alkaloids with narcotic properties similar to nicotine are present in mountain laurel's leaves, flowers, and seeds. The quinolizidine alkaloids begin to show their deleterious effects within an hour after

Sophora secundiflora, mountain laurel; mature seedpod dehiscing to show red seeds.

ingestion, leading to violent, bloody vomiting, headache, dizziness, fever, extreme thirst, and breathing difficulties, eventually culminating in convulsions and death. Some animals do consume the seeds without fatal consequences, probably because the seeds are encased in a very hard seed coat, which allows the potentially fatal gift to pass unopened through the digestive tract.

Despite the potent poisons contained in the seeds, they were used by several tribes in vision-seeking rituals during the Red Bean Dance. The seeds were sometimes part of a "medicine bundle," a pouch containing sacred objects, as has been documented by archaeologists who have unearthed such talismans. Today some tribes still use the seeds to decorate clothing or articles associated with ceremonies or hunting.

The Comanche used the seeds to treat earache, and the Cheyenne used them to make an eyewash.

The tree is toxic to cattle, sheep, and goats that consume the foliage, which becomes more toxic with age. The seeds are even more poisonous than the leaves.

Mountain laurel is a popular landscaping tree throughout the Southwest. The flowers' nectar attracts honeybees.

GERANIACEAE (Geranium Family)

Erodium cicutarium

STORKSBILL, RED-STEMMED FILAREE, CUT-LEAF FILAREE

The long, pointed, birdlike beak on the seedpod of *Erodium* species is acknowledged in the Greek word *erodios*, meaning "heron," and is also the basis of the common names "storksbill" and "cranesbill" (see p. 313). The geranium family name emphasizes this structure as well, with *geranos* meaning "crane." The very finely divided, fernlike leaves give rise to storksbill's species name, *cicutarium*, referring to their resemblance to the leaves of *Cicuta*, or poison hemlock (see p. 264).

Storksbill arrived in America in the 1700s with Spanish explorers and has become naturalized throughout North America, including New Mexico, where it colonizes disturbed sites from 3,000 to 7,500 feet elevation. This low-growing annual overwinters as a rosette. Its tiny, purple, five-petaled flowers are among the very first to appear in the spring and continue blooming until August and beyond. The delicate blooms soon form vertical seedpods shaped like a stork's bill, each containing five seeds with a corkscrew appendage. As the pod ruptures, the appendages flip the seeds away from the mother and then respond to changes in humidity by twisting, eventually drilling the seeds into the earth.

Storksbill leaves, containing 17 percent protein and high amounts of minerals, may be eaten raw or as a cooked vegetable. The cooked taproot tastes somewhat like a turnip.

Storksbill has diuretic, astringent, and anti-inflammatory properties. Native American women used it to relieve water retention and heavy bleeding associated with their menstrual periods or childbirth. New mothers of the Jemez tribe ate storksbill to increase their milk supply.

Erodium cicutarium, storksbill; seedpods.

It also has a history of use for diarrhea, urinary tract infections, and wounds. A tea brewed from storksbill leaves made a good face wash for acne. The leaves were also used in a hot bath for the relief of arthritis pain.

The dried plant was mixed with stored watermelon seeds to keep them insect-free. The Isleta and Ramah Navajo value storksbill as nutritious early spring forage for their livestock. Storksbill is also food for a wide assortment of wildlife, including desert tortoises, bighorn sheep, deer, and rodents. The seeds are eaten by harvester ants, rodents, and birds.

LAMIACEAE (Mint Family)

Mentha arvensis

FIELD MINT

In Greek mythology, Mentha was a naiad nymph of freshwater streams who had the misfortune to be infatuated with Pluto. When Queen Persephone, Pluto's wife, realized that Mentha was being taken advantage of by her husband, she blamed Mentha and transformed her into a plant fit only for treading on. Why she didn't zap her husband into a cocklebur for preying on the young maiden remains an open question. Pluto softened Mentha's cruel fate by making her an eternally sweet, fragrant, and useful plant.

Field mint is found in northern and central New Mexico around streams and in moist wooded areas up to about 9,000 feet elevation. The

three-foot-tall plant forms extensive colonies by means of its underground rhizomes. Its blue, or sometimes pink or white, flowers bloom from July to September. Like all other members of the mint family, its paired leaves are opposite each other on the square stems.

Also like other members of the mint family, the leaves are the repository of the volatile oils that give each species its characteristic scent. The glands in the leaves of field mint produce fragrant essential oils that lend their distinguishing taste to beverages, jellies, and sauces. These oils are also important in the perfume and pharmaceutical industries.

Mentha arvensis is the primary source of menthol, which stimulates a cooling sensation when ingested (think of spearmint chewing gum), inhaled (disguising the bitterness of nicotine in cigarettes), or put on the skin to soothe sore muscles (think "Biofreeze!"). Menthol has local anesthetic properties that work by blocking sodium channels, thus reducing neural activity and helping to bring about pain relief. The major commercial production of *M. arvensis* for menthol extraction is in Uttar Pradesh, India, but there is also some important cultivation in the Pacific Northwest.

Another chemical isolated from field mint is thymol, used extensively in cough syrups and known also for its antibacterial and antifungal qualities, as well as its pleasant fragrance. In addition to killing microbes, it will rid the body of parasitic worms. Thymol is currently being studied for its antitumor activity.

New Mexico's native peoples have used field mint for food and medicine for centuries. The Kayenta Navajo flavored meat and mush with it, and they also made a lotion to reduce swelling. The Ramah Navajo, Acoma, and Laguna made a mint tea to treat fever, headache, and other symptoms of colds and flu. The Isleta made a mint poultice for eye problems. Early European settlers used field mint for pain relief as well, particularly rheumatism.

Field mint is very easy to grow in a home garden and does well in shady areas.

LAMIACEAE (Mint Family)

Monarda pectinata

PAGODA PLANT, PONY BEEBALM

The genus name, *Monarda*, honors the Spanish physician and botanist Nicolas Monardes, who lived an extraordinarily productive ninety-five years in the 1500s, during which he wrote a three-volume work on the medicinal properties of New World plants. His treatise was so popular that there were forty-two editions translated into six languages. The species name, *pectinata*, means "comblike."

The design of beebalm's flower structure is striking, with clusters of flowers at intervals along the stems, resembling a pagoda or a shish kebab, depending on whether you prefer architectural or gustatory imagery.

Pagoda plant can be an annual or a perennial and grows to three feet tall. Flowering from early spring through August, it grows in northern and western New Mexico, in dry sandy soil in grassland to ponderosa pine communities from 4,000 to 8,000 feet elevation.

Like other *Monarda* species, pagoda plant has a number of unique volatile compounds in the oils in its leaves, which are important in the fragrance industry and are also excellent insect repellents. The principal source of the aroma and the beneficial effects is thymol, an aromatic oil that kills bacteria, fungi, and parasitic worms, particularly hookworms.

People have long enjoyed pagoda plant's leaves in a refreshingly aromatic herbal tea, which has the added benefit of fighting off infections and soothing a sore throat. All *Monarda* species are considered to be carminative, a cleverly polite way to say that they are antiflatulents, or intestinal gas reducers. The crushed plant has also been used in liniments as a counter-irritant for painful joints and muscles. The Navajo enjoyed the pleasant aroma of dried pagoda plant hanging in their hogans. Native American women used this plant to relieve menstrual pain.

Pagoda plant's value as a distinctive vertical element in a landscape or wildflower garden should not be underestimated. It is also attractive in dried flower arrangements. Plants can be propagated by cuttings as well as by seed, which is widely available commercially.

Livestock generally avoid pagoda plant. However, hummingbirds, bees, and butterflies, including dull firetip (*Apyrrothrix araxes*) and pine white (*Neophasia menapia*) butterflies, are frequent visitors.

Also pictured here is *Monarda punctata*, which resembles pagoda plant and has similar chemical constituents.

Monarda punctata, spotted beebalm; whole plant.

LAMIACEAE (Mint Family)

Poliomintha incana

ROSEMARY MINT, PURPLE SAGE, FROSTED MINT

Rosemary mint is a three- to four-foot-tall shrub with a grayish appearance due to its dense covering of hairs. Its scientific name emphasizes the gray aspect: *polios* is Greek for "whitish gray," and *incana* means about the same thing in Latin. We can therefore add "gray-gray mint" to its list of common names!

Rosemary mint grows in gypsum soil in the Chihuahuan Desert and Four Corners area of New Mexico from 4,000 to 6,000 feet elevation. It is a common plant at White Sands National Monument, where this photo was taken. Blooms are present from May through October and are delicate lavender to white; the deeper purple calyx gives the illusion that they are somewhat darker.

The pleasant aroma of rosemary mint hints that the flavor is just as appealing. Indeed, the Tewa and Hopi ate the leaves raw or as a cooked vegetable, and they dried the plant for winter sustenance. A refreshing herbal tea can be brewed from the leaves, and the flowers can be used for seasoning.

Comanche medicine women kept rosemary mint leaves in their mouths while they chewed other medicinal plants to camouflage the unpleasant taste and increase the efficacy of the other herbs (an early use of the adjuvant concept!). The Hopi and Tewa used rosemary mint to treat rheumatism and ear trouble. The Navajo used it to heal sores and also smoked it like tobacco in ceremonial offerings to eagles.

Rosemary mint is worthy of more extensive use in xeriscapes and natural landscape plantings.

RANUNCULACEAE (Buttercup Family)

Delphinium wootonii

WOOTON'S LARKSPUR

Larkspur's graceful profile, complete with a long floral spur, resembles that of a dolphin, as reflected by its genus name, *Delphinium*, meaning "dolphin." The species name honors Elmer Wooton, a twentieth-century American botanist who established the New Mexico State University herbarium in 1890.

Wooton's larkspur is a tall perennial that grows throughout New Mexico in caliche grasslands, desert scrub, and oak woods up to 6,000 feet elevation. It has a brief bloom period, from April to May.

All parts of larkspur contain poisonous alkaloids and may be fatal to livestock, especially cattle and particularly in the spring, when concentrations of the toxins are at their highest. The poisons become less dangerous as the plant matures. The toxins primarily affect the neuromuscular system, causing falling and difficulty breathing, followed by death in just a few hours.

Delphinium wootonii, Wooton's larkspur; seedpods.

Larkspur seeds were used by early medical practitioners to induce vomiting and purging and to treat asthma and congestive heart failure. The most common use of larkspur preparations was as a parasiticide to remove lice and scabies mites.

SOLANACEAE (Nightshade Family)

Quincula lobata

PURPLE GROUND CHERRY

The significance of purple ground cherry's genus name is obscure, although *quin*, the Latin root for "five," may refer to the five fused petals. The species name, *lobata*, clearly references the deeply lobed leaves.

Purple ground cherry is a plant of the Chihuahuan Desert and is found throughout eastern New Mexico. It blooms from early spring until frost, growing in sandy or caliche soil in fields or woodlands.

This perennial grows very close to the ground. A close inspection reveals that the buds and the margins of the mature leaves have a healthy covering of sticky glandular hairs, and thus the plant tends to have an untidy appearance, with dirt collected on its leaves.

Native Americans made jelly and salsa from the ripe berries. (Note that the immature fruits may be toxic.) They also had several medicinal uses for the plant. They crushed the roots to put on wounds and used the leaves for a tea to remedy coughs. A warm compress of the ripe fruits was applied as a poultice for sore throat.

Purple ground cherry is usually viewed as a weed, but it is neither aggressive enough nor ugly enough to be classified as such. In fact, it's sufficiently charming to be a ground cover or border plant in a xeriscape, where its bright blooms and strong drought tolerance especially shine.

SOLANACEAE (Nightshade Family)

Solanum elaeagnifolium

SILVERLEAF NIGHTSHADE, HORSENETTLE

The pain-relieving aspect of *Solanum*'s narcotic properties is extolled in the genus name, probably derived from *solamen*, Latin for "solace." Hundreds of years before Linnaeus named this genus in the 1750s, the group was known as "nightshade" for the plants' soporific (sleep-inducing) qualities, sometimes drawing the shades of night on a permanent basis. The silvery color and to some extent the shape of the leaves are similar to those of the genus *Elaeagnus*, hence *elaeagnifolium*, or "*Elaeagnus* leaves."

Silverleaf nightshade is found throughout New Mexico in disturbed soil in grasslands from 3,500 to 6,500 feet elevation. The plants may be covered with prickles in desert areas where drought prevails or have few to no prickles where rainfall is somewhat higher. It establishes large colonies, propagating by rhizomes, which are extremely difficult to eradicate. To a wildflower lover, the large purple blooms are gorgeous from March to October, but farmers have an entirely different perspective, as these plants can quickly overtake a cultivated field.

Silverleaf nightshade berries have their own remarkable story. Although the berries contain the poisonous glycoalkaloid narcotic solanine, they are also rich in enzymes, which the Navajo, Zuñi, Cochiti, and Pima add to boiling milk as a rennet to separate it into solid curds for cheese and liquid whey for a drink. Hispanics use the berries in a similar process to make *asadero*, a delicious white cheese made for home use as well as commercial sale.

Solanum elaeagnifolium, silverleaf nightshade; mature berries.

The berries, especially fresh green ones, should never be eaten raw because of the solanine toxin, which can cause vomiting, stomach pain, dizziness, headache, heart arrhythmia, and, finally, shades of night.

The toxic glycoalkaloids have been investigated for their effectiveness in limiting the growth of certain types of cancer and in controlling the herpes virus. The roots have antiseptic properties, as well as the ability to draw fluid from swollen tissue.

Native Americans used silverleaf nightshade as a remedy for assorted ailments. Zuñi medicine men chewed the roots prior to sucking the venom from a rattlesnake bite, then placed the macerated roots on the wound. The Zuñi also put a piece of root into a tooth cavity to dull the pain.

Solanum elaeagnifolium is a source of naturally occurring steroids, including solasodine and diosgenin, which are used by the pharmaceutical industry in India to produce birth control hormones. It is interesting to note that Keresan women drank a tea made from silverleaf nightshade berries to sustain the flow of milk after childbirth, in effect using it as an early form of oral contraception.

The digestive enzymes contained in the berries made them effective, when crushed and mixed with animal brains, in tanning hides.

VERBENACEAE (Verbena Family)

Glandularia bipinnatifida (syn. *Verbena bipinnatifida*)

PRAIRIE VERBENA, DAKOTA VERVAIN, MOCK VERVAIN

Prairie verbena's charming gingerbread-man-shaped flowers remain an endearingly beautiful attraction in the Southwest. Its currently accepted genus name, *Glandularia,* apparently refers to glands on the seeds. The species name recognizes the deeply lobed leaves, which appear to be "twice pinnate," or *bipinnatifida.*

Prairie verbena is a low-growing, hairy-leaved perennial found throughout New Mexico in grassland, pinyon-juniper, and ponderosa pine communities from 2,800 to 8,000 feet elevation. It blooms from March to October, covered in flowers despite the heat or drought.

Prairie verbena is a Life Medicine for the Navajo. The Acoma and

Glandularia bipinnatifida, prairie verbena; inflorescence gone to seed.

Glandularia bipinnatifida, prairie verbena; whole plant.

Laguna crushed the leaves with rocks and rubbed them on snakebites. They also steeped the leaves in hot water to make a gargle for sore throat.

This plant is a must-have in xeriscape or natural landscapes as a border plant or ground cover. Its purple to pink flowers attract a wide assortment of butterflies. The plants will form mats that smother weeds and help stabilize the soil. The leaves have an "eau de bovine" cowpen smell that repels sheep and deer.

VERBENACEAE (Verbena Family)

Verbena macdougalii

NEW MEXICAN VERVAIN, MACDOUGAL VERBENA

Verbena is an ancient Latin name for vervains in Europe, where their leaves and branches were considered sacred and were used for effecting

Verbena stricta, slender vervain; scenic view.

magical cures. They came to be known as holy herbs because of a legend purporting that Jesus's wounds were stanched with this plant after He was taken down from the cross.

The species name of this stately New Mexico vervain honors Daniel MacDougal, a nineteenth- and twentieth-century American plant physiologist who became a principal American authority on desert ecology.

The three-foot-tall perennial is found in central and western New Mexico, growing in open areas of grasslands and along roads up to about 8,000 feet elevation. The coarsely hairy leaves are in pairs along very hairy square stems. The flowers bloom from June to October in progressively upward rings along the spikes, with the encircling band of open flowers resembling the tonsured hairstyle of medieval monks. The flowers are pollinated by flies and bees.

MacDougal verbena can be brewed as a diuretic tea, which also can reduce fever and relax muscles. The Navajo used the tea as a ceremonial lotion, and they burned the leaves as a fumigant during certain rituals.

Also pictured here is a purple cousin, *Verbena stricta,* a two-foot-tall annual that grows in central New Mexico, in sandy or caliche soil in prairies and along roadsides. It flowers from July to September and attracts butterflies and bees. The seeds are consumed by small mammals and birds. Livestock will not eat this plant, so an abundance of it in a pasture indicates poor grazing management. It is used medicinally in the same way as MacDougal verbena.

Pink Flowers

APOCYNACEAE (Dogbane Family)

Apocynum androsaemifolium

SPREADING DOGBANE

To fully appreciate dogbane, one must hold in balance its pretty flowers and fragrance on the one hand and its toxic personality on the other.

If the neighbor's dog is bothering your garden, a few judiciously placed dogbane plants might be in order. The genus name says they will keep the "dog" (*kyon*) "away" (*apo*). Some "catbane" would be helpful, too.

Open dry sites, meadows, and riparian zones near streams from 5,000 to 10,000 feet elevation are equally suitable for dogbane. It recovers well from fires because it propagates so rapidly by rhizomes.

Like other members of this family, dogbane has a milky sap containing poisonous glycosides, primarily apocynamarin, which acts on smooth muscles such as the heart and uterus. Even in tiny doses, the active ingredient constricts blood vessels and raises blood pressure, accounting for its use by Native Americans to help expel the placenta after childbirth, stop heart palpitations, and control nosebleeds. These root decoctions were also used for congestive heart failure, convulsions, urinary problems, earache, and sore throat.

Juice from the fresh roots was used to treat syphilis. The Sanpoil people in what is now Washington State used a weekly dose of dogbane root as a contraceptive. Dogbane served as an early Viagra for desperate Okanagan-Colville men, who chewed and swallowed fresh dogbane leaves or smoked dried leaves as an aphrodisiac. Others applied dogbane sap to a bald head to stimulate hair growth, while the Cherokee used a root decoction to treat mange in dogs.

The stems were soaked to release the fine silky fibers, which were used for sewing as well as for making twine, fishing nets, hunting nets, storage bags, fabric, and strings for bows.

Bees harvest the nectar for honey, and monarch butterflies (*Danaus plexippus*) use the plant as a larval host.

APOCYNACEAE (Dogbane Family)

(formerly in Asclepiadaceae)

Asclepias speciosa

SHOWY MILKWEED, BUTTERFLY-WEED, COMMON MILKWEED

This four-foot-tall, handsome plant with distinctive pink flowers is aptly named "showy," which is also the meaning of its species name, *speciosa*.

Showy milkweed is a perennial, growing at higher elevations (5,000 to 8,500 feet) and in more water-favored habitats in mountain woodlands and along streams than its cousins. It blooms from June through August. With successful pollination, the flowers will transform into seedpods covered with tubercular bumps.

All milkweeds are poisonous, but some are more so than others. Showy milkweed is one of the lesser evils, making it less risky for consumption and just as efficacious as a medicine. Many Southwest tribes ate showy milkweed after cooking it in water to remove the bitter alkaloids. They cooked meat with the buds and flowers, which tenderized the meat with the protein-digesting enzyme asclepain and thickened the broth.

Showy milkweed has a long history of medicinal use by western and northwestern tribes. The roots were the most common source of the active ingredient and were used to treat stomachache, venereal diseases, headache, rheumatism, tubercular cough, and swelling. The latex was rubbed on skin sores and cuts as an antimicrobial and a healing balm. The latex was also used as a facial beauty treatment, a wart and callus remover, and a ringworm cure. There was surely a fine line between the beauty effects and burning: the sap was also used as a temporary brand for livestock!

The dried seedpods made handy spoons or mini-dippers. The fluff on the end of the seeds was used to make baby diapers. It was also used by early American colonists as mattress stuffing and to fill life vests during World War II when imports of kapok were halted. Milkweed species were tested by the USDA in the 1980s as a substitute for imported goose down. Yields were found to be between 200 and 300 pounds per acre. About 125 pods would be needed to produce a pound of down.

Bees and other hymenopterans are important pollinators, and butterflies use showy milkweed as a nectar source. Caterpillars of monarch (*Danaus plexippus*), queen (*Danaus gilippus*), and soldier (*Danaus eresimus*) butterflies graze on showy milkweed and other *Asclepias* species.

For information about other milkweed species, see pp. 44 and 266.

ASTERACEAE (Sunflower Family)

Carduus nutans

NODDING THISTLE, MUSK THISTLE, BRISTLE THISTLE

"Thistle nodding" is the literal interpretation of the Latin name *Carduus nutans,* which Linnaeus conferred in the mid-1700s. Its pink, bristly, shaving-brush-like flower heads, consisting only of disk flowers, are thrust well above the leaves like periscopes and impart a faint aroma of musk. Beyond these pleasant features, the rest of the plant has no cuddliness to endear it to either gardeners or farmers. The leaves are vigorously spiny, arming four- to six-foot-tall stems, which have "wings" running their length.

As a biennial, the young plant spends its first year in a rosette close to the ground, then bolts, or shoots up a flower stalk, in the second year. The pollen is too sticky to be dispersed by the wind, so honeybees, bumblebees, and various other long- and short-tongued nectar imbibers effect pollination. Seeds are produced in abundance and are responsible for the plant's rapid spread in freshly disturbed areas, such as after a fire, along roadsides, or in overgrazed land.

In its native regions of western and central Europe, *Carduus* and its cousin *Cirsium* have been used for medicine for more than two thousand years, and during the Middle Ages they became known as "holy thistles" because their healing properties were ascribed to saints. However, when they arrived in North America in the 1850s, mixed in hay for livestock, holy thistles became wholly troublesome. More than 150 years later, nodding thistle is found in forty-five states, including New Mexico, where it is classified as a Class B noxious weed, meaning that it is a nonnative plant that is more than well represented in some areas and should be contained so that it does not spread to other areas.

If one can get past all the spines, the peeled stems are tasty when cooked as a vegetable. The large, heavily armed flower buds can also be cooked and eaten like artichokes.

Thistles have been subjected to numerous classification reshuffles, from *Carduus* to *Cirsium* and back again, then split and reassigned. Meanwhile, *Carduus nutans* continues its inexorable spread across the country, oblivious to the taxonomic angst its classifiers have suffered.

ASTERACEAE (Sunflower Family)

Centaurea americana (syn. *Plectocephalus americanus*)

AMERICAN BASKET FLOWER, AMERICAN STAR THISTLE, POWDERPUFF

In Greek mythology, Chiron, one of the centaurs (half-man, half-horse creatures), was wounded in the foot by Hercules but was cured with a plant in this genus. The synonymous genus name comes from the Greek *plektos*, meaning "twisted" or "plaited," and *kephale*, meaning "head," in reference to the tangled appearance of the disk flowers when the flower heads are past their prime. *Americana* indicates that this species is "made in America."

A key feature of this plant gives it one of its common names, "basket flower." The phyllaries, or bracts under the flower head, are delicately

Centaurea americana, American basket flower; field view.

fringed and overlap to form a basket (see main photo).

This slender annual reaches five feet tall, usually with a single stem that is branched at the top and adorned with showy pink and creamy white flowers that are up to five inches in diameter. Seeing a hillside in bloom with these eye-catching saucers in May through July makes you immediately start planning how to get seed to plant in your own garden. The flower head contains only disk flowers, most of which are cream-colored and fertile, but the outer ones, which are gender neutral, are frilly pink and longer than the rest, creating the illusion of ray flowers (see photo).

American basket flower can grow in caliche soil but performs most prolifically in moist sandy or loamy soil such as it might find in a garden

Centaurea americana, American basket flower; flower head in full bloom, with drooping outer disk flowers.

setting. It is found throughout southern New Mexico to just north of Interstate 40, growing in grasslands, fields, and roadsides up to 7,000 feet elevation.

The Kiowa of the southern plains made a poultice of basket flower leaves to treat skin sores.

Basket flower is easy to cultivate, requires little water, and tolerates some shade. Not only is it a sweet-scented attraction for butterflies, bees, and other insects, but it is also a beautiful cut flower.

A related species, very similar in appearance, is *Centaurea rothrockii* (syn. *Plectocephalus rothrockii*), which grows at higher elevations in the ponderosa pine zone.

For information about related species, see pp. 49 and 185.

ASTERACEAE (Sunflower Family)

Chaenactis douglasii

PINK THIMBLE, DOUGLAS'S DUSTYMAIDEN, MORNING BRIDES

A close look at pink thimble's flower head reveals that it has only disk flowers. The genus name, *Chaenactis*, means "open rays" in Greek, reflecting the plant's outer disk flowers, which are flared open like a circle of gaping baby bird mouths, giving the impression of being ray flowers. The species name honors David Douglas, a Scot who, though he lived only thirty-five years, made extensive botanical collections throughout the American West, earning him great respect in the Royal Horticultural Society.

Dustymaiden flowers from June to August, growing along roadsides, on hillsides, and in disturbed soil from 3,000 to 11,000 feet elevation. The flower heads and their pedicels are covered in tiny, sticky glands. The appearance of this species is not always consistent, probably because the number of chromosome sets varies according to location and type of soil.

The Gosiute of Utah crushed the plant and used it as a poultice for sore muscles, swelling, and sprains and as a treatment for snakebite. Pink thimble tea was used to treat coughs, slow a racing heart, and relieve headache.

Dustymaiden is a successful pioneer species in disturbed land such as roadsides and mining sites. It is even competitive against invasive weeds such as knapweed and cheatgrass that are troublesome in the arid Southwest. Commercial production of its seed has made it readily available for restoring wildland areas, and it participates with other species in improving wildlife habitat and providing food for the animals. It also attracts multitudes of insects, making it an especially important buffet table for sage grouse during rearing of their young.

Chaenactis douglasii, pink thimble; seed head.

ASTERACEAE (Sunflower Family)

Cosmos parviflorus

WILD COSMOS, SOUTHWESTERN COSMOS

Cosmos parviflorus literally means an "ornament with small flowers."

Its dainty flowers, just over an inch in diameter, with pink, white, or lavender ray petals, are borne on branched two-foot-tall plants. The delicate, threadlike leaves give the plant an overall graceful appearance that is attractive in a garden setting.

The bloom period is from June through October. The seeds are achenes with two barbs, with which it hitches a ride on fur or clothing. It grows in central and southwest New Mexico at elevations up to about 10,000 feet.

Cosmos parviflorus, wild cosmos; seed head.

The Ramah Navajo use a cold tea made from this species as a lotion during ceremonial chants.

Livestock find wild cosmos palatable, so it is found mainly in areas protected from grazing and sometimes in disturbed sites such as roadsides.

ASTERACEAE (Sunflower Family)

Palafoxia sphacelata

SAND PALAFOXIA

The genus name, *Palafoxia*, was apparently given to this plant in 1816 in honor of Juan de Palafox y Mendoza, a bishop who founded the University of Mexico in the early 1600s. However, that commemoration was not properly documented. Meanwhile, in the early 1800s, another

Palafoxia sphacelata, sand palafoxia; field view.

Palafoxia sphacelata, sand palafoxia; seed heads with papery pappi.

Palafox, José Rebolledo de Palafox y Melzi, was enjoying an illustrious career as archbishop of Mexico, viceroy of New Spain, and more. When a subsequent Spanish botanical treatise was published in 1858, the genus name *Palafoxia* was attributed to this new national hero, upstaging the previous Palafox for whom the name was originally given. Today, more than two hundred years later, this nomenclatural attribution has yet to be resolved.

The species name, *sphacelata,* means "withered, appearing to be dead." That description is incongruous with the lovely appearance of the plant. As the common name indicates, sand palafoxia grows in dry sandy or caliche soil, from 3,000 to 6,000 feet elevation. The flowers bloom from June through October and produce abundant seeds.

The phyllaries and upper stems are covered in purple glands, making the plant sticky to touch. The glands also catch dirt and other detritus, making it look scruffy and unkempt. The papery pappi atop each seed (see photo) resemble those of *Uropappus lindleyi* (see p. 81).

Sand palafoxia has medicinal value for treating fever, chills, and nausea.

Bring this plant into a native landscape to add color and texture and to welcome the many bees, birds, and butterflies that use it as a nectar source.

ASTERACEAE (Sunflower Family)

Stephanomeria tenuifolia (syn. *Lygodesmia tenuifolia*)

WIRELETTUCE, NARROW-LEAF STEPHANOMERIA

The feathery bristles on the achenes of wirelettuce are the basis of the Greek genus name, *Stephanomeria*, meaning "wreath division," apparently referring to the plumose pappus, with each strand looking like a soft, white mascara brush. In some regions of the Southwest, *Stephanomeria tenuifolia* plants closely resemble those of a cousin, *S. pauciflora*, and some taxonomists believe they should be grouped under a single species name.

Stephanomeria tenuifolia, wirelettuce; seed head.

The plant grows in low desert, grassland, and pinyon-juniper communities from 3,000 to 7,000 feet elevation. To the casual observer, it looks like a round heap of skinny little green bones. Instead of the leaves doing the food production work, it's these wiry, leafless stems that carry on the photosynthetic process. In June through September, these "bones" come to life with dainty pink flowers decorated by four notches at the petal tips. The flower head contains only ray flowers, typical of the chicory tribe of the sunflower family. These flowers are rapidly replaced with plumose white pappus bristles that will wing the seeds away into new territory.

For the Navajo, wirelettuce is a Life Medicine. They used the roots as a narcotic and to make a strong tea drunk to stimulate delivery of the placenta during childbirth. They also collected the dried, milky sap and chewed it like chewing gum.

BIGNONIACEAE (Catalpa Family)

Chilopsis linearis

DESERT WILLOW, DESERT CATALPA, MIMBRE

The genus name highlights the "liplike" (*cheilos*, "lip," and *opsis*, "similar to") floral features of both the calyx and the corolla. The bilabiate (two-lipped) calyx features a pair of puckered lips, while the corolla's lower lip has three lobes and protrudes like that of a sulking child.

The species name speaks of the "narrow" (*linearis*) leaves that can be up to a foot long, as well as the long seedpods, which are about the same length and width as the leaves.

The common name "desert willow" draws attention to its resemblance to the familiar willow trees, with their long leaves, supple branches, and attractive shape. The Spanish name "mimbre" means "willow tree" and is also the word for "wicker," indicating its use in making baskets or furniture.

Chilopsis linearis, desert willow; close-up showing bilabiate corolla and calyx.

Desert willow is a monotypic species, meaning that it is the only member of its genus. The flowers are very similar to those of a relative, *Catalpa speciosa*, but the leaves of the latter are deltoid and far wider. Desert willow is found in the southern half of New Mexico, as far north as Albuquerque, while *C. speciosa* is found in northern New Mexico.

Desert willow grows in desert washes and arroyos from 2,800 to 6,200 feet elevation and can tolerate alkaline soil. Like mesquite and Apache-plume (see pp. 104 and 334), it is a phreatophyte, a tree or shrub whose deep roots put it in constant contact with an underground water supply. This small tree can reach thirty feet tall, and its slender trunk and branches bear a relaxed, open crown. It is usually found in well-drained limestone soil but will grow well in almost any soil type as long as it is not overwatered.

The fragrant flowers are in large clusters that bloom in "flushes," with the tree covered in blooms at intervals throughout the summer. The seedpods are two-celled capsules that split lengthwise to release seeds that have fringed white hairs on both ends. The split pods remain on the dormant tree through the winter and into the next season alongside the flowers and fresh green pods.

The flowers and pods are edible, and a tea made from the flowers has been used to relieve coughs. Desert willow contains compounds with antifungal properties; the bark and the wood contain lapachol, which has shown anti-inflammatory efficacy in preclinical and clinical studies and has potential use in reducing cancer metastasis.

Desert willow has some use as fence posts and firewood, and Native Americans used it to make coils for baskets, hunting bows, baby cradles, and frames for houses and grain storage units. The Cahuilla of Southern California used the bark to make clothing and nets.

Desert willow leaves contain bitter cyanogenic glycosides, which react when chewed to release hydrogen cynanide, a poison. Not surprisingly, this tree is not on the menu for livestock or other ruminants. Various bird species, however, eat the seeds, and bees and hummingbirds are frequent visitors to the flowers. Desert willow provides nesting sites for a variety of birds and cover for other wildlife.

Desert willow is useful in land reclamation and erosion control, helping to stabilize soil along waterways and highways. Several varieties are available, with flower colors ranging from deep purple-pink to white. It grows from seed but is most easily propagated by cuttings. Nursery stock is readily available for transplanting.

This tree is a ubiquitous feature of gardens and xeriscapes in the Southwest because of its graceful form, gorgeous flowers, sweet fragrance, longevity, moderate deer resistance, and general ease of upkeep.

Chilopsis linearis is a caterpillar host for the geometer moth (*Eucaterva variaria*) and the rustic sphinx moth (*Manduca rustica*). Adult monarch butterflies (*Danaus plexippus*) feed on the flowers.

CACTACEAE (Cactus Family)

Cylindropuntia imbricata (syn. *Opuntia imbricata*)

CHOLLA

Cholla (pronounced "choya") has "cylindrical" stems, hence the opening note in the genus name, *Cylindro*, which is combined with *Opuntia*, the name given by Philip Miller in the 1700s to the cactus group. The species name, *imbricata*, means "overlapping."

Cylindropuntia imbricata, cholla; field view.

A "forest" of cholla in bloom across the grasslands is an astounding vision of bright wine-colored flowers, short branching trunks, and multitudes of thorns. Mature plants are often taller than a person and grow in a variety of soil types in desert, grassland, and pinyon-juniper communities up to about 5,000 feet elevation. Blooming from May to August, the flowers are succeeded by bright yellow fruits (see photo) that persist through the fall and into the next flowering season. A dominant species in the Chihuahuan Desert, cholla displays variation in spine characteristics and height throughout its range.

Cylindropuntia imbricata, cholla; fruit.

The Acoma and Laguna roasted cholla stems for food and also dried them to store for the winter. The Tewa ate the boiled fruits with cornmeal porridge. The Papago ate cholla fresh in the spring and summer. Modern Puebloans eat cholla both stewed and raw, and they grind the dried fruits into flour. Most Native Americans have traditionally enjoyed eating the flower buds in the spring; in fact, March has been called "the cactus moon" due to the scarcity of other food sources.

The Acoma and Laguna removed the cholla needle coverings and ground them into a paste to use on boils. The men of these two tribes ate the needle coverings to make them tough in times of war. These tribes also scraped the pith out of the center of the stems and used it to treat earache. The Hopi used cholla roots mixed with globemallow as a remedy for diarrhea, and they made a tea to rinse their hair and make it shiny.

The thorns were used for sewing needles and tattooing. The juice from the crushed stems contains polysaccharides that made it a good fabric stiffener. The long, dried stem skeletons, with their decorative patterns of holes, were used for candles and torches, and nowadays are shaped into walking canes or unusual frames for flower arrangements.

Fresh canes can be planted side by side to grow into a living fence, in the same way *Fouquieria* (ocotillo) (p. 14) can be used.

Cholla is a must-have feature in a waterwise landscape. The flowers attract a wide assortment of insects in pursuit of its pollen and nectar. The well-armed plants provide a good "security fence" for nesting birds such as the cactus wren and offer strong deer resistance.

Cholla is an indicator plant for ruin mounds, where human activity altered the soil pH, texture, and nutrient content to favor the plants.

For information about *Cylindropuntia* species, see p. 300.

CACTACEAE (Cactus Family)

Echinocactus texensis

HORSE CRIPPLER CACTUS, DEVIL'S HEAD, CANDY CACTUS

Horse crippler cactus, covered with dangerous spines and lurking in camouflage among grasses, can wreck the hooves of a horse, the tires of a vehicle, or the soles of your shoes. *Echinocactus*'s very robust spines, noted by the Greek word for "spiny," *echino*, are mostly flat and curved with

Echinocactus texensis, horse crippler; close-up of spines.

horizontal ridges, are positioned in clusters of six to eight per areole, and cover the hemispheric globe that emerges about six inches from ground level.

This cactus is found in the Chihuahuan Desert of New Mexico, mainly in the southeast counties, growing in rocky limestone soil from 3,000 to 4,000 feet elevation. The large, frilly pink flowers that bloom in the spring transform into decorative red Easter egg–like fruits.

Example of cactus thorn fishhook tied with cordage. Artifact found in cave on Lower Pecos River.

Horse crippler cactus can be strategically included in a natural or waterwise landscape not only for its handsome flowers but also for its attraction to insects, especially bees. Not surprisingly, it is also deer resistant. The fruits draw both avian diners and tortoises. People in Mexico make cactus candy and a chocolate-colored cheese with these fruits. Native Americans heated and bent the long spines to use as fishhooks (see photo).

CACTACEAE (Cactus Family)

Echinocereus reichenbachii

LACE CACTUS, PINK-FLOWERED HEDGEHOG CACTUS

Lace cactus's genus name means "spiny-waxy," describing the plant and the surface of the flowers. The species name honors Heinrich Reichenbach, a nineteenth-century German botanist who was an inter-

Echinocereus reichenbachii, lace cactus; head cactus cheerleader: "Yea for cactus!"

nationally recognized orchidologist and the director of the botanical garden at the University of Hamburg.

The six- to eight-inch-tall lace cactus grows in the Chihuahuan Desert, in grasslands and oak-juniper woodlands up to 4,500 feet elevation. The small, bristly spines, borne atop the tubercular protuberances on the stem's surface, form an orderly phalanx of lacy spider legs over the plant's interface with the world. Each cluster has sixteen to twenty spines spreading at right angles in a circle. The buds and necks of the flowers are covered in a cobweb of long, tomentose hairs. Spectacular frilly pink flowers appear in May and June.

This cactus has great merit in xeriscape plantings because of its flowers, attraction to birds and insects, and deer resistance.

For information about another *Echinocereus* species, see p. 13.

CAPRIFOLIACEAE (Honeysuckle Family)

Symphoricarpos rotundifolius

MOUNTAIN SNOWBERRY, ROUNDLEAF SNOWBERRY

The white berries of *Symphoricarpos rotundifolius* are recognized with the common name "mountain snowberry." The species name, *rotundifolius*, describes the plant as being "oval-leaved."

Mountain snowberry is a three- to five-foot-tall shrub with long, narrow, bell-shaped pink to white flowers that mature into small, fleshy white fruits known as drupes. It grows throughout New Mexico, except in the east, from 7,000 to 10,000 feet elevation.

The white berries are high in saponins, steroidal glycoside compounds that have soapy properties, making the berries undesirable as a food but effective for medicinal as well as for cleaning purposes. The purgative property of the berries and leaves was employed by some Native Americans to purify their bodies prior to certain ceremonies. The Kayenta Navajo used the berries to treat sore throat.

Mountain snowberry is widespread and abundant within its range and is a dominant species in subalpine and alpine communities, where its leaves serve as forage for wildlife. Its saponin content reduces its palatability, and the nutrient content is not stellar, but as an early riser from the long sleep of winter, it appeals to both wildlife and livestock, which seek out its green shoots. Palatability, which is based on the saponin content, varies according to the location, the plant part, and the season. Grouse, pheasant, and magpies consume the berries.

The plants provide cover for many small mammals and birds.

Mountain snowberry reproduces vegetatively by rhizomes and from basal sprouts at the root crown, as well as by seed, making it valuable for reclaiming damaged land and for establishing plant cover on roadsides after construction. Seed is available commercially for use in reclamation and for planting as an ornamental.

CLEOMACEAE (Cleome Family) (formerly in Capparidaceae)

Cleome serrulata

ROCKY MOUNTAIN BEE PLANT, SPIDERFLOWER, PINK CLEOME

Bee plant's dainty pink flowers and extravagantly long stamens waving in the breeze in desert landscapes demand that you stop for a closer look. *Cleome*, a name from antiquity for a mustard plant, was resurrected by Linnaeus for this species. *Serrulata* means "with fine teeth."

Rocky Mountain bee plant is a three-foot-tall, somewhat bushy annual with long racemes of flowers that bloom from July to September. The stamens are exserted well past the petals, giving it a "spidery" appearance. The leaves are palmate, with three leaflets (see photo), and when they are crushed, they give off an unpleasant smell. Bee plant grows in northern and western New Mexico, in disturbed areas from 4,500 to 8,000 feet elevation.

New Mexico's ancient inhabitants ate the seeds, which archaeologists have found in coprolites and burial sites. Today bee plant is an indicator of Indian ruin mounds.

The Acoma and Laguna gathered the seeds, cooked them, and then spread them out to dry for winter use. The Isleta ground the seeds into flour for making bread; others ground them into a meal for mush. The present-day Hopi cultivate bee plant in their fields to ensure a source of greens in the spring and seeds later on.

High in iron, calcium, and vitamin A, bee plant leaves have been called "Indian spinach." The Navajo boil the greens in at least two changes of water before eating them. They also enjoy the leaves mixed with onion and meat in a stew. Other Native Americans grind the seeds together with corn to make bread, or even chop the entire plant to mix with cornmeal to make cleome cornbread.

Cleome serrulata, Rocky Mountain bee plant; close-up of flower with common red soldier beetle (*Rhagonycha fulva*), which preys on other insects in the flower.

Cleome serrulata, Rocky Mountain bee plant; palmate leaves.

Cleome serrulata, Rocky Mountain bee plant; mature seedpods.

The people of the San Ildefonso and Santa Clara Pueblos made a leaf tea to treat stomach problems. The warm leaves also were used to poultice sores and wounds. The Navajo used cleome to treat ant bites, and they still use it as both food and medicine for the Night Way Ceremony. Somewhat ironically, they made a shoe deodorant from the smelly leaves—fighting fire with fire!

The plant's high iron content made it valuable as a pigment for painting pottery. The leaves were boiled and then pressed down and dried into bricks. This substance, called "guaco," could later be reconstituted with water and used to draw black or gray designs on pottery prior to firing. A yellow-green dye for wool is made by soaking the yarn for two weeks in fermenting *Cleome* leaves. The flowers yield a lilac dye.

Bee plant is recommended as a garden ornamental not only for its fanciful appearance but also because it hosts a multitude of insect species, including bees and butterflies that feed on its nectar and pollen. It is also useful for stabilizing disturbed or damaged land and for highway beautification projects. Doves and other small birds eat the seeds.

CONVOLVULACEAE (Morning-glory Family)

Convolvulus arvensis

FIELD BINDWEED, POSSESSION VINE, GLORYBIND

Field bindweed's dainty "fairy dancer" flowers belie the pernicious evil lurking in its heart. Bindweed is one of the worst agricultural weeds in the world, causing great economic loss in many crops, especially in the temperate regions of the world. The genus name tells us the basis of its wickedness: *convolvere* is a Latin word meaning "to entwine." Bindweed wraps itself around crop plants or any other handy plants, smothering them into weakness or death. Its species name is based on the Latin *ensis*, "place of origin or habitat," and *arvens*, "field," which is where bindweed wreaks its havoc.

Since its first introduction into the United States in the early 1700s, probably in contaminated seed, it has been reintroduced at various times, which has helped hasten its spread throughout this country and Canada. In New Mexico, it is classified as a Class C noxious weed, meaning that it

Convolvulus arvensis, field bindweed; leaves.

is already totally out of control and landowners are encouraged to do all they can to protect themselves and others from it.

This perennial has a vast root system that extends twenty feet deep and stores massive amounts of carbohydrates, which help it regenerate when the top is cut back. The seeds are equally tough, living up to fifty years in the soil. Blooming from May to July, the flowers are pink to white, with five petals fused to form a funnel about an inch wide. The plant grows in any disturbed soil, dry or moist, but is not found in very wet environments.

Aside from appropriate herbicides designed to control bindweed, other "natural" means of killing it include intense grazing by sheep and uprooting by pigs, which apparently rate bindweed rhizomes high on their list of favorite foods. Bindweed has a number of insect predators that may have potential as a means of biological control.

The Ramah Navajo made an infusion of bindweed to treat spider bites; they drank the tea and also rubbed it on the bite. Bindweed is also used as a purgative. It promotes wound healing and acts as an anti-

Convolvulus equitans, dagger bindweed; close-up of flower.

Convolvulus equitans, dagger bindweed; seedpods.

inflammatory. The women of some California tribes drank bindweed tea to control excessive menstruation.

The flowers produce abundant nectar and attract a wide range of pollinators, including honeybees, bumblebees, halictid bees, and butterflies. Bindweed is a larval host for painted crescent (*Phyciodes picta*) and Mexican fritillary (*Euptoieta hegesia*) butterflies.

Field bindweed has a homegrown cousin, *Convolvulus equitans*, which has a more congenial and much less aggressive personality. The leaves are more narrow (compare photos), and the flowers form a tighter funnel, with an outward curving tip at each corolla union. The flowers of this species come in two color forms: one is white, and the other, the one no doubt preferred by New Mexico State University fans, is white with a maroon center.

CONVOLVULACEAE (Morning-glory Family)

Ipomoea leptophylla

BUSH MORNING-GLORY

In *Ipomoea*, the stems twist and turn as they grasp other plants or objects to twine around and climb, and the petals in the bud spiral just like an old-fashioned ice cream cone. *Ipomoea* means "like a worm," referring to this spiraling behavior and giving a new perspective on Shakespeare's phrase "as the worm turns." The species name, *leptophylla*, describes the

Ipomoea leptophylla, bush morning-glory; bud with spiraling corolla.

Ipomoea leptophylla, bush morning-glory; seedpods.

plant's "thin leaves." But instead of climbing like most *Ipomoea* species, the stems of bush morning-glory form arching mounds as they grow.

Bush morning-glory is a robust perennial found mainly in the northern half and eastern plains of New Mexico, growing in sandy soil in the grasslands and hills. The flowers put on their show from May through July, opening in the morning and retiring for a siesta by noon. Bush morning-glory is easy to grow from seed and makes an attractive feature in landscapes and revegetation projects.

The enormous root is edible, but the Great Plains tribes would eat it only when other food was not available. They did, however, use it for medicine. The Lakota ate the root when they had stomach problems. The Pawnee dusted the dried, powdered root over the body for pain relief, and they inhaled the smoke from the burning root to calm themselves when they were nervous or suffering from bad dreams. The root was also used in treating urinary tract problems.

To ensure that their adult horses were big and strong, the Keresan peoples fed the huge root to young colts to help them grow and to mares to make them fertile.

Ipomoea leptophylla, bush morning-glory; field view.

FABACEAE (Bean Family)

Calliandra eriophylla

CALLIANDRA, FAIRY DUSTER

When a flower presents the viewer with inch-long, hot pink stamens, you just know the name has to acknowledge that feature. Indeed, Calliandra is from the Greek words meaning "beautiful stamens." The actual flower is only one-fifth inch in size, but somehow those long stamens manage to be very compactly arranged within the bloom until the moment is right for them to stretch to their full glorious length. When all twenty or so of them are fanned out, they do indeed look like a duster for fairies. The species name, *eriophylla* (wool leaf), refers to the fine white hairs covering the immature leaflets on the bipinnately compound leaves.

This three-foot-tall shrub blooms from February through April and may flower again in the fall. The seedpods are flat and become reddish brown, splitting to eject the seeds. Calliandra grows in southwest New Mexico in desert scrub and grassland communities from 4,000 to 5,000 feet elevation.

Plants propagate not only by seed but also by rhizomes. Calliandra shrubs are very popular as landscape features, and they attract butterflies, bees, flies, and hummingbirds. The plant is a larval host for the statira sulphur butterfly (*Aphrissa statira*). Deer and javelinas browse the leaves, and doves, finches, and quail eat the seeds.

FABACEAE (Bean Family)

Dalea formosa

FEATHER DALEA, FEATHERPLUME

The genus name was given in honor of Samuel Dale, a seventeenth- and eighteenth-century British physician and botanist who wrote a book on medicinal plants. *Formosa* means "finely formed" or "beautiful."

Feather dalea is a small, rigid shrub that grows about three feet tall and has tiny pinnate leaves. From April to June, the shrub is covered in splendid little rose-colored flowers, each with a yellow detached banner accenting the rose petals like a tiny yellow neckerchief. A distinctive

Dalea formosa, feather dalea; note small pinnately compound leaves.

Dalea frutescens, black dalea; close-up of flower.

feature is the long, feathery lobes of the calyx. After the petals have fallen off, the hairy calyces remain, adding to the ethereal featheriness of the plant.

Feather dalea grows throughout New Mexico, except in the Four Corners area, in rocky caliche or sandy soil on hills and slopes in plant communities up to 7,000 feet elevation.

A sweet-flavored tea made from the flowering branches is enjoyable for its own sake or as a home remedy for colds and flu. The Apache used the tea to assuage bone pain. By boiling down the leaves, the Jemez, Acoma, and Laguna produced a cathartic for cleansing the digestive tract and for increasing endurance in long-distance runners.

Feather dalea is a xeriscape delight, valued for its visual effects; its attraction to butterflies, bees, and other insect life; and its very undemanding maintenance requirements.

Also shown here is a relative, *Dalea frutescens*, a shrub found in the southern part of New Mexico that has become popular as a feature in xeriscapes.

For information about other *Dalea* species, see pp. 195 and 308.

NYCTAGINACEAE (Four O'clock Family)

Allionia incarnata

TRAILING FOUR O'CLOCK, TRAILING WINDMILLS

The genus name honors the eighteenth-century botanist Carlo Allioni. The species name is Latin for "flesh pink," which seems too pale a color to represent four o'clock's deep pink flowers. The plant's stems are prostrate, trailing along the ground for up to ten feet, as the common names indicate.

This perennial inhabits southern and western New Mexico, growing in sandy soil in disturbed areas of desert and grassland communities below 6,000 feet elevation. The stickiness of its hairs leads to a slightly unkempt appearance, because dirt and debris adhere to its surfaces.

The leaves are paired along the stems, with one larger than the other—a characteristic of the four o'clock family. Blooming from April through October, the "flowers" are actually clusters of three flowers, whose close union creates the impression of a single, symmetrical unit. Furthermore, the colorful "petals" are actually sepals, which can be distinguished by the number of vascular bundles (veins): petals have one, while sepals have three.

Trailing four o'clock is a potent hallucinogen, useful for Native Americans on a vision quest as part of their healing ceremonies. On a more down-to-earth level, the Ramah Navajo steep the root in water, then use the liquid to wash a swollen body part. In the past, trailing four o'clock tea was a remedy for diarrhea and diseased kidneys.

NYCTAGINACEAE (Four O'clock Family)

Mirabilis multiflora

DESERT FOUR O'CLOCK, MARAVILLA, COLORADO FOUR O'CLOCK

Mirabilis multiflora, meaning "wonderful" and "many flowers," decorates central and western New Mexico from April to September with clouds of magenta pink. The flowers open in the afternoon, as the family name implies, and remain open through the night.

Desert four o'clock has dense foliage and a rounded profile that is about four feet wide and two feet tall. It grows in sandy or rocky soil in desert grassland to pinyon-juniper communities up to 7,500 feet elevation.

Several members of the four o'clock family have exceptionally large roots, and *M. multiflora* has a giant one, at one foot in circumference and twice as long. It has long been used as a source of food and medicine. Archaeologists discovered large quantities of roots in the two-thousand-year-old Fresnal Shelter in New Mexico's Sacramento Mountains. The pith of the root was collected for a range of medicinal purposes. The Navajo treated inflammation and rheumatism with it, and Puebloans consumed it after overeating—something like a modern-day Tums or Alka-Seltzer!

The Zuñi used the root, cooked into bread, as an appetite suppressant to encourage weight loss. Today modern *curanderas* recommend chewing it for the same purpose. The people of the Santa Clara Pueblo used the root to make a wash for infected eyes and a rubbing compound to soothe sore muscles. Contemporary *curanderas* prepare a similar treatment by mixing the powdered root with petroleum jelly or olive oil and applying it to inflamed joints.

Hopi medicine men used desert four o'clock to induce visions when they were seeking spiritual guidance in making difficult diagnoses. The Navajo, known for their weaving skills, still value this plant as a source of light brown or purple dye for wool.

Desert four o'clock is a striking member of a xeriscape planting or natural landscape and can serve as an extensive ground cover. It can tolerate some shade. Plants can be grown from commercially available seed, cuttings, or root division.

The flowers are pollinated by hawk moths such as the great ash sphinx (*Sphinx chersis*) and Achemon sphinx (*Eumorpha achemon*), as well as by various bee species and hummingbirds.

OROBANCHACEAE (Broomrape Family)

(formerly in Scrophulariaceae)

Castilleja sessiliflora

PAINTBRUSH, DOWNY PAINTED-CUP

The genus name, *Castilleja*, honors Domingo Castillejo, a late-eighteenth-century professor of botany in Spain. The species name indicates that the flowers are sessile, or attached directly at the stem.

Castilleja spp., paintbrush; seedpods.

Castilleja sessiliflora flowers resemble a shrimp emerging from the colorful bracts, which may be pink, yellow, or purplish. The bracts are not part of the flower, but are simply modified leaves.

Paintbrush's bright palette of bract colors is behind the myth about the origin of this plant. In the legend, paintbrushes in olden days were green leafy plants that a Native American artist dipped into his paints to depict the sunset on a deerskin canvas. As he finished with his "paintbrushes," he tossed them over his shoulder, and where they landed, they grew, with their leaves still stained by the paints.

Castilleja sessiliflora grows to almost two feet tall throughout New Mexico, except in the Four Corners area. It is usually found in sandy soil, in many plant communities from 4,500 to 7,000 feet elevation.

Some wildlife and livestock, including elk, deer, and sheep, use paintbrush as a minor part of their diet. Sage grouse eat the flowers and fruits. Bumblebees are the primary pollinators for *C. sessiliflora*, with its unusual flower shape precluding most insects from reaching the interior.

For information about other paintbrush species, see p. 18.

PLANTAGINACEAE (Plantain Family)

(formerly in Scrophulariaceae)

Penstemon ambiguus

PINK PLAINS PENSTEMON, SAND PENSTEMON

The first part of the scientific name, *Penstemon*, accurately describes the plant's "five stamens," of which one is a staminode, or a sterile stamen, hence *ambiguus*, "doubtful."

Pink plains penstemon in bloom is a splendid mound of color. Usually less than two feet tall, this perennial grows in sandy soil in grasslands up to about 6,000 feet elevation. It blooms from May through October. Propagation is by seed and by spreading from the woody crown at ground level.

Pink plains penstemon contains bitter and toxic iridoid alkaloids that give it a range of pharmacological properties, including antiviral, anti-

Penstemon ambiguus, pink plains penstemon; immature seedpods with "tail" of remaining style.

Penstemon ambiguus, pink plains penstemon; with black blister beetle (*Epicauta pennsylvanica*).

Penstemon crandallii, Crandall's beardtongue.

spasmodic, and antitumor effects. One of these, verbascoside, is an anti-inflammatory compound relevant to the Navajo's use of the plant to treat wounds and burns. An infusion of pink plains penstemon was used as an emetic (vomit inducer) by the Acoma and Laguna in purification rituals.

The Hopi and Tewa timed their watermelon planting with the beginning of pink plains penstemon's blooming season.

This plant is a must-have in a xeriscape or natural landscape, for its masses of pink blooms, which also attract a wide assortment of insect life (see photo). Note that it grows rather slowly and takes several years to realize its full potential of blooming splendor. It is also useful for stabilizing sandy soil, to which it is well adapted. Seed is commercially available.

Penstemon palmeri, Palmer's penstemon.

Two other charming pink-flowered penstemons include *Penstemon crandallii* (Crandall's beardtongue), a low-growing species, and *P. palmeri* (Palmer's penstemon), a tall, handsome, fragrant plant with buds that look like the fat little pink cheeks of a Cabbage Patch doll (see photos).

For information about other Penstemon species, see pp. 20 and 169.

POLEMONIACEAE (Phlox Family)

Phlox nana

SANTA FE PHLOX

This uniquely charming blaze of pink is waiting to join your xeriscape. The name *Phlox nana* (small flame) speaks of the bright color in a small package. The starburst pattern of white in the center of the bright pink flower is guaranteed to catch your eye. The stems and leaves are covered with sticky hairs that trap sand and give the plant an unkempt look.

Santa Fe phlox grows throughout central New Mexico in dry sandy soil on rocky slopes in desert shrubland to juniper-pinyon woodlands from 3,500 to 5,000 feet elevation. With a robust taproot, this tiny perennial is well equipped to withstand drought and heat while blooming continuously from April to September. It reseeds and also spreads slowly from rhizomes. The traits that make it so appealing in the wild make it equally attractive in a landscape setting.

Early Spanish settlers used the petals to treat fever blisters and extracted a poor-quality red dye from the plant.

ROSACEAE (Rose Family)

Erythrocoma triflora (syn. *Geum triflorum*)

PRAIRIE SMOKE, OLD MAN'S WHISKERS, THREE-FLOWERED AVENS

Imagine a mountain meadow full of pink-haired troll dolls from the 1960s, and you will immediately recognize prairie smoke. The recently bestowed genus name *Erythrocoma* means "red hair," in reference to the

Erythrocoma triflora, old man's whiskers; note pink bracts and sepals covering barely visible pale petals.

feathery styles that grow long and fluffy as the seeds mature. Each flower stalk usually has "three flowers" (*triflorum*) nodding from the top.

This two-foot-tall perennial has very hairy, basal, pinnately compound leaves. It is found in northern and western New Mexico on hillsides and meadows in ponderosa pine and aspen communities up to 9,000 feet elevation, usually in moist conditions. The flowers have pinkish to purplish sepals and bracts, which almost completely obscure the greenish-white petals. Prairie smoke propagates by both seed and rhizomes.

Native Americans in the Pacific Northwest made a beverage from the roots and crushed seedpods that was said to taste like sassafras tea. Not only was it drunk for its pleasant flavor, but it was also used as a remedy for sore throat and as a soothing wash for irritated eyes or for a nursing mother's sore nipples.

In Canada, Okanagan-Colville women made a love potion from the roots to win the attentions of a man.

ROSACEAE (Rose Family)

Rosa woodsii

WILD ROSE, WOODS' ROSE

Rosa is an ancient Latin name for a rose. The species name honors Joseph Woods, an eighteenth- and nineteenth-century Englishman who wrote a popular treatise on British roses.

Rosa woodsii, wild rose; fruits (rose hips).

Wild rose is a tangled shrub that can reach ten feet tall. It has pinnately compound leaves, hooked thorns, and striking, fragrant pink flowers. It usually grows near a water source in pinyon-juniper to ponderosa pine and aspen communities from 4,000 to 10,000 feet elevation. Wild rose is both a colonizer of disturbed sites and a member of climax communities.

The flowers bloom from May to July, followed by red fruits, called "rose hips," which remain on the plant over the winter or until consumed by wildlife or humans. The plants can be five years old before they bloom.

Ripe, red rose hips, like those shown in the photo, have long been valued as one of the richest sources of natural vitamin C and are still sold in health food stores today. Rose hip tea, jam, jelly, and syrup continue to be popular, as are the dried fruits, which are used in fruitcakes and puddings. During the food shortages in Britain during World War II, rose hips replaced citrus fruits as a source of vitamin C to prevent scurvy.

Native American women and children collected the fruits to eat fresh as snacks and to dry for later use. The people of many tribes, including the Apache and Navajo, ate the fruits raw or boiled them to make jelly. Isleta women fed their babies rose petal tea before their first taste of mother's milk.

Pueblo people along the Rio Grande used wild roses to treat stomach pain, toothache, wounds, and sore throat. The Navajo and Kawaiisu used the strong, flexible stems of rosebushes for baskets. The Navajo fashioned the limbs into the hoops used in the Holy Way Ceremony. Rose thorns provided needles for leatherwork. Pioneer women steeped rose petals in water to use as perfume.

Wild rose leaves provide good forage, and rosebushes provide cover for small and large mammals, as well as for birds. Wild rose has a vigorous rhizome system that makes it very effective in erosion control and in revegetating disturbed soil. The plants are also ideal for landscaping purposes, with their beautiful flowers and ability to attract birds and other wildlife. Wild rose hosts the mourning cloak butterfly (*Nymphalis antiopa*).

SAPINDACEAE (Soapberry Family)

Ungnadia speciosa

NEW MEXICO BUCKEYE, MEXICAN BUCKEYE, TEXAS BUCKEYE

New Mexico buckeye's genus name recognizes Baron Ferdinand von Ungnad, ambassador of the Austrian emperor Rudolf II in the 1570s. The species name, *speciosa* (showy), is very apropos. New Mexico buckeye is indeed a handsome small tree, about twelve feet tall, with long leaves divided into paired leaflets. It is found mainly in southern New Mexico on rocky slopes, as in this photo taken at Organ Mountains Recreation Area. Flowering season is March and April.

Unfortunately, this lovely little tree has a nasty secret: the foliage and seeds contain toxic glycosides that are poisonous to birds, mammals, and even fish. Fishermen can make an easy catch by crushing the seeds to release the toxin and then throwing them into the water to stun the fish. Native Americans used the poison on the tips of their arrows to enhance their killing power. The shiny black seeds are sometimes carried for good luck, and they are also used for necklace beads. Children play with them like marbles. People who eat the seeds experience dizziness, nausea, and abdominal discomfort, and large doses can be fatal.

Ungnadia speciosa, New Mexico buckeye; empty seedpods.

The toxins do not prevent bees from producing excellent honey from the flowers, however. With its showy, fragrant flowers, this tree is very attractive for shade and in landscapes. It has the added benefit of being deer resistant. Butterflies and moths enjoy its nectar, and larvae of the Henry's elfin butterfly (*Callophrys henrici*) feed on the foliage.

SARCOBATACEAE (Greasewood Family)

(formerly in Chenopodiaceae)

Sarcobatus vermiculatus

GREASEWOOD, BLACK GREASEWOOD

Greasewood is a very common plant in the alkaline soils of the Southwest, but because of its nondescript appearance, it garners little attention. The genus name, *Sarcobatus*, alludes to the "fleshy" (*sarco*) leaves and the spiny, "brambly" (*batos*) growth habit. The species name, *vermiculatus*, means "wormy," possibly referring to the thick, green, wormlike leaves.

Greasewood is a ten-foot-tall shrub found throughout New Mexico, especially in the western and northern regions, up to 8,000 feet elevation. It flowers from spring through late fall and is usually monoecious, with the male flowers spatially separated from the females on the same plant and normally blooming before the females, which are shown in the main photo. Greasewood often grows in heavy-textured saline soils in

Sarcobatus vermiculatus, greasewood; staminate (male) flower.

floodplains, dry arroyos, and gullies where the water table can be reached with its deep taproot.

The original human inhabitants of New Mexico ate greasewood seeds only in hard times. They could be made palatable by roasting. The Keresan tribes, as well as the Navajo, crushed greasewood leaves to treat insect bites. The Paiute burned the plant and mixed it with water to use as a medicine for diarrhea and rectal bleeding.

Greasewood branches are remarkably straight (see photo), and the wood is quite strong. The Navajo carve images of snakes from the wood to use in ceremonies such as the Lightning Chant, Beauty Chant, and Mountain Chant. They also use the durable wood for planting sticks, knitting needles, dice, and weaving tools. The Hopi use the wood in construction and for rabbit sticks, musical rasps, and arrows. For the Hopi, the wood is one of the four prescribed fuels used in kiva ceremonies.

When a Navajo girl has her coming-of-age ceremony, one of her treasured gifts is a bundle of greasewood stirring sticks (see photo), which she will use for cooking throughout her life, perhaps even bequeathing

Sarcobatus vermiculatus, greasewood; whole plant. Note straight stems, used for tools.

Navajo stirring sticks.

them to her descendants. They will hang in her kitchen in an honored place and within easy reach. The smooth, strong sticks are considered to be "women's weapons," a counterpoint to the men's arrows, because women use these tools to protect their families from hunger, poverty, and disease by cooking healthful foods for them.

Livestock and wildlife feed on greasewood, which is an important winter browse for them. Although they enjoy its salty taste, the high concentration of oxalates can cause poisoning if the animals consume large amounts of it in a short time. It is also important as cover for many wildlife species.

Greasewood is useful in landscape settings, as well as in stabilizing soil and revegetating mining areas or saline sites where other species cannot grow.

White Flowers

AMARANTHACEAE (Amaranth Family)

(formerly in Chenopodiaceae)

Salsola kali subsp. *tragus* (syn. *Salsola tragus*)

TUMBLEWEED, RUSSIAN THISTLE

Salsola, from the Latin *salsus* (salty), is a halophyte, or salt lover, that makes its home in deserts with soil containing salty compounds formed from sodium and potassium. The species name, *kali*, is derived from the Latin word for "potassium."

This icon of western cowboy movies made its debut in the United States as a stowaway in a shipment of grain from Russia in 1877. People in Eurasia, as well as in the southwestern United States, did not let its disreputable character as an agricultural pest deter them from making use of its alkaline properties for commercial and home-use products. Since

Salsola kali subsp. *tragus*, tumbleweed; edible seedling.

Salsola kali subsp. *tragus*, tumbleweed; field view.

ancient times in the Middle East and Mediterranean region, *Salsola kali* plants were burned to produce soda ash, which was used in making glass and soap.

The Ramah Navajo used an infusion of tumbleweed ashes both internally and externally to treat smallpox and flu. The Navajo chewed tumbleweed leaves to make a poultice for treating insect bites. Tumbleweed contains salsolin, which acts as a blood pressure regulator, constricts blood vessels, and causes uterine muscle contractions. A liquid extract of the plant serves as a diuretic. It is also an antihelminthic, or dewormer.

Looking at the long, soft, spider-leg-like leaves of a baby *Salsola*, as shown in the photo, one would never suspect that it would grow up to be a bristly, prickly, rolling tumbleweed tough enough to dent a car in a windstorm. The Navajo eat the seedlings of tumbleweed as a tasty green vegetable, consuming it either fresh or cooked with small pieces of mutton fat. In the past, they also collected the teensy-weensy seeds, which contain 40 percent protein, for roasting.

Tumbleweed is a minor food source for both large and small mammals as well as birds. However, it may be toxic to livestock because it can accumulate as much as 5 percent oxalic acid. Nitrates may accumulate to a level that is toxic to sheep.

ANACARDIACEAE (Sumac Family)

Rhus microphylla

LITTLE-LEAF SUMAC, DESERT SUMAC

The genus name, *Rhus*, is based on the Greek word for "sumac," *rhous*. Little-leaf sumac has pinnately compound leaves with five to nine "very small leaflets," or *microphylla*. The stemlike structure to which these leaflets are attached is called a rachis, which, as seen in the photo, has "wings."

Little-leaf sumac is polygamodioecious, a quick way of saying that plants have either male or female flowers (dioecious), but that they also have quite a number of flowers on any given plant that are "perfect," with both male and female parts in a single flower. The flowers appear between March and May before the leaves and are borne in clusters at the end of branches, where they will mature into orange-red fruits covered with tiny fuzz during July and August (see photo).

Rhus microphylla, little-leaf sumac; fruit.

It is as tough as its cousin *Rhus aromatica* subsp. *trilobata* (common name "skunkbush"), also pictured here, in tolerating heat, drought, cold, and poor soil conditions. Little-leaf sumac does not produce seeds as abundantly as skunkbush, nor is its fall wardrobe as splashily brilliant. Its habitat stretches from 3,500 to 6,000 feet elevation, while

Rhus aromatica subsp. *trilobata*, skunkbush; fruit.

skunkbush grows up to 7,500 feet, covering desert grassland to pinyon-juniper zones.

The Apache ate the zingy, tart fruits either raw or cooked into jam. Several Indian tribes macerated the fruits and soaked them in hot or cold water to make a lemonade-like drink (sumacade!).

Native Americans and modern-day herbalists use little-leaf sumac for its significant amount of tannins, which have astringent properties that constrict soft tissue and stop bleeding. Drinking a tea made from the leaves and chewing the fruits can alleviate discomforts of the common cold.

The roots of little-leaf sumac yield a yellow dye. Little-leaf sumac's dainty foliage, bright red fruits, and hardy growth habit make it a valuable xeriscape plant that can be pruned into a hedge. Its ability to grow in almost any kind of soil and to resprout from roots and rhizomes makes it useful in stabilizing soil in degraded or disturbed areas. However, be aware that at certain times of the year, it may look less than lovely, because little-leaf sumac hosts caterpillars of the Chinati sheepmoth (*Hemileuca chinatiensis*), which feed on the leaves and may defoliate it. The dense, interwoven branches create good cover for small mammals and birds. Mule deer, pronghorn, and occasionally livestock will browse the plant, but the palatability is poor. Various birds and small mammals consume the tangy fruits.

APIACEAE (Carrot or Parsley Family) (formerly in Umbelliferae)

Cicuta maculata

WATER HEMLOCK, SPOTTED PARSLEY, SPOTTED COWBANE

The genus name, *Cicuta*, is the ancient Latin name for poison hemlock. The stem's purple pigmented areas are acknowledged by the species name, *maculata*, which is Latin for "stain" or "spot."

The leaves of water hemlock consist of long, narrow, toothed leaflets. The entire leaf is well over a foot long. The plant reaches about five feet tall, with ridged, purple-streaked stems. The tiny white flowers bloom from June to September and are arranged in clusters called umbels, which are about five inches in diameter. Water hemlock grows in wet places, whether along marshes or on stream banks, in central New Mexico.

Water hemlock is native to nearly all of North America and is considered to be one of the most toxic plants in the world, alongside castor bean. Do not taste test any member of the carrot family in the wild

Cicuta maculata, water hemlock; field view.

unless you are truly an expert in distinguishing one from another. Only a small amount passing your lips is likely to be fatal. The agent of death is cicutoxin, which is present in all parts of the plant at all stages of growth. The roots hold the greatest concentration of this toxin, especially in the early spring. The effects on the central nervous system include convulsions, salivation, and grinding of teeth, followed by death in one to eight hours, for both humans and livestock. In the past, the roots were mixed with rattlesnake venom or decomposed liver to make poison arrows.

The Paiute treated swollen muscles and joints, as well as rattlesnake bites, with a poultice made from water hemlock's roots. This remedy not only reduced inflammation but also deadened the pain. The Cree treated rheumatism with a liniment made from the powdered roots. Others used root poultices to cure lameness, running sores, and cuts. In a stretch of faith or of desperation, the Cherokee used the roots for contraception, eating a piece of root for four days in a row to achieve permanent sterilization.

Water hemlock exhibits insecticidal properties. Indian tribes soaked corn seed in an extract of the roots before planting to ward off insect pests. Black swallowtail (*Papilio polyxenes*) caterpillars feed on water hemlock, undeterred by the toxins.

APIACEAE (Carrot or Parsley Family) (formerly in Umbelliferae)

Heracleum maximum

COW PARSNIP, COW CABBAGE, HERCULES-PARSNIP

This largest species of the carrot family in North America was named for Hercules, the Roman demigod famous for his great strength and size. Hercules is also said to have used this plant for medicine. The species

name, *maximum*, also invokes the giant stature of cow parsnip, whose leaves can reach one foot across on eight-foot-tall stems.

Cow parsnip blooms from May to August, depending on the elevation, with the flowers organized in large, beautiful umbels that may be up to eight inches across. It prefers rich soil in wet riparian areas and is a higher-elevation plant within the upper pinyon-juniper zone (7,500 feet) to the ponderosa pine and spruce-fir zone (9,000 feet).

It was a popular food in the springtime, when Native Americans peeled the flower stalks and ate them with the tender and sweet aromatic leaves. They also ate the cooked roots, which have a taste similar to turnips. The stems of cow parsnip are hairy and can cause a skin rash if brushed against or handled when preparing them for eating. Note that the stems and leaves should be consumed with caution, because the stems contain furanocoumarin, which can cause light sensitivity.

Cow parsnip's medical prowess was tapped by more than fifty North American Indian tribes, who used all of its parts for medicine. The roots were used for a wide assortment of medical ills. Cooked and mashed roots were used in a poultice for rheumatism, wound healing, bladder ailments, and sore throat. Head cold sufferers breathed the smoke from a mixture of cow parsnip roots and the sap of *Pinus monophylla* (single-leaf pinyon). A tea made from the roots was used to halt diarrhea or, over a longer treatment period, cure tuberculosis and syphilis. Some tribes used the raw root as an anodyne for tooth pain, inserting pieces of it into cavities. Powdered roots were rubbed onto the gums when teeth became loose. The roots were also used in puberty rituals and in purification ceremonies for warriors and hunters.

Cow parsnip is palatable to livestock and other herbivores, including elk and bears. It is a larval host and nectar source for short-tailed swallowtail (*Papilio brevicauda*), black swallowtail (*Papilio polyxenes*), and anise swallowtail (*Papilio zelicaon*) butterflies.

APOCYNACEAE (Dogbane Family)

(formerly in Asclepiadaceae)

Asclepias asperula

ANTELOPE HORNS, SPIDER MILKWEED

In Greek mythology, Asklepios was the god of medicine and healing, and his human acolyte was Aesculapius, a leading medical practitioner of his time. Linnaeus honored both of them by naming this group of milkweeds *Asclepias*, in recognition of the plants' healing properties. The

species name, *asperula*, means "a little rough," probably in reference to the knobby texture of the seedpods.

Antelope horns is a perennial, growing throughout New Mexico in limestone soils in grasslands and pinyon-juniper communities up to about 6,000 feet elevation.

The active ingredients in *Asclepias* species, found in all parts of the plant, are potent alkaloid glycosides, which act on smooth muscles, such as the heart and uterus, and on the nervous system. The Acoma and Laguna made a tea from *Asclepias asperula* to treat fever and cardiac problems, and the Tewa made a tea from the roots to quell chest pains. The

Asclepias asperula, antelope horns; inflorescence.

Navajo and Isleta inhaled powdered leaves and stems to clear congestion and a stuffy nose. The Navajo also used a leaf tea as an emetic to clean out the digestive system.

Antelope horns is toxic to all livestock, particularly sheep, with the poisonous compounds desglucosyrioside and syrioside primarily responsible for the ill effects. About 1 percent of the animal's body weight in green plant material (about one pound for a one-hundred-pound animal) is enough to wreak havoc on its health—or more to the point, its life. Within a few hours of eating antelope horns, the animal will suffer acutely and die.

One person's poison is another person's elixir, and monarch butterflies make good use of these toxic compounds against their avian predators. *Asclepias asperula* and its cousin *A. viridis* are primary food sources for monarch (*Danaus plexippus*), queen (*Danaus gilippus*), and soldier (*Danaus eresimus*) butterflies. The adults consume the nectar, and their caterpillars feed exclusively on *Asclepias* leaves. As the caterpillars eat, they are also ingesting the alkaloids. These alkaloids are stored away and then transferred into the adult as the larva passes through the pupa stage. Birds that feed on monarchs pay a price: if they eat one or two, they will vomit; if they gobble down several, it's death bells. Birds retain this discriminatory knowledge and learn to focus on more healthy food.

Asclepias stems provide strong, soft fibers, which the Tewa and Zuñi used to make string and rope (see photo). The downy tufts (coma) on the ends of the seeds were spun into yarn and woven into fabric.

Example of soft cordage. Artifact found in cave on Lower Pecos River.

Asclepias subverticillata, horsetail milkweed; inflorescence.

Asclepias subverticillata, horsetail milkweed; seed with coma attached.

Asclepias subverticillata, horsetail milkweed; seedpods.

Antelope horns, with its low growth habit and high drawing power for butterflies, is an attractive species in a xeriscape or natural landscape. It is also resistant to deer.

The narrow-leaved and whorled species, such as *A. subverticillata* (shown here), are considered the most dangerous because their glycoside alkaloids act on the nervous system rather than the cardiovascular system. Ingesting only 0.2 percent of body weight in green leafy material can be fatal.

For information about other milkweed species, see pp. 44 and 215.

ASPARAGACEAE (Asparagus Family) (formerly in Agavaceae)

Dasylirion wheeleri

SOTOL, DESERT SPOON

Sotol's genus name, *Dasylirion*, is Greek for "bushy lily." The species name honors George Wheeler, a nineteenth-century army engineer who directed fourteen surveys of the western United States.

The very young flower head of *Dasylirion wheeleri* can be fermented into a "white lightning" called sotol, in much the same way that mescal or tequila is made from *Agave*. Before the Spanish arrived in America, sotol had the alcohol content of beer, but with the introduction of distillation techniques, the proof level, and the kick, went up. Now sotol is an internationally recognized beverage.

Sotol grows in southern and west-central New Mexico, in desert grasslands from 3,000 to 6,000 feet elevation. It flowers from May to August, with flower stalks reaching up to fifteen feet. The leaves have a wide, fleshy "spoon" where they attach to the stems. These can be cooked and eaten in much the same way as artichoke bracts.

Dasylirion wheeleri, sotol; female inflorescence.

Sotol is dioecious, meaning that males and females are on separate plants. The flower stalks are generally similar in appearance from a distance. Up close, however, females are distinguishable by their three-winged seeds (see photo) and males by their little, pinkie-length catkin clusters.

The central stem, or "heart," of sotol is sweet. Southwest Indians baked the crown, the zone above the roots where the leaves attach, then formed the starchy material into patties, which they left out to dry. These sweet, chewy cakes were eaten immediately or stored for later use. The Apache and Papago roasted or boiled the young, tender flower stalks to eat as a vegetable. The Mescalero cooked the heart with bones to make soup.

Ancient peoples baked sotol and yucca in earth ovens. These structures are associated with burned rock middens, which are common in early habitation sites in the Chihuahuan and Sonoran Deserts, with charred remains of sotol and yucca still present in them. Sotol seeds were also a popular food, and evidence of their consumption before AD 1000 has been found in feces in Lower Pecos Canyonlands archaeological sites.

The limber leaves have strong fibers that are used to make cordage, baskets, head rings, mats, and sandals (see photos). In the past, the dried

Example of sandal fragment. Artifact found in cave on Lower Pecos River.

Example of basket fragment. Artifact found in cave on Lower Pecos River.

flower stalks were used to make frames for cradleboards and headdresses for Mountain Spirit dancers, as well as for fences, roofs, and walking sticks.

Southwest ranchers utilize the sweet trunks and leaf bases of sotol to feed cattle during drought. Bighorn sheep and javelinas use sotol as browse.

Sotol is a conspicuous feature in a xeriscape, with its wide circle of thin leaves and tall, narrow flower stalks. It grows well from seed and is easy to propagate from the shoots sprouting from around its crown.

ASPARAGACEAE (Asparagus Family) (formerly in Agavaceae)

Yucca baccata

BANANA YUCCA, DATIL

Yucca is the state flower of New Mexico, and *Yucca baccata* is one of its key representatives.

Banana yucca has the widest range of all *Yucca* species, growing throughout New Mexico, except the far eastern regions, in a range of soil types in desert grassland to pinyon-juniper communities from 2,800 to 8,000 feet elevation. The flowers are in dense panicles, or clusters, that do not rise much, if at all, above the leaves. The leaves are concave and stiff, up to forty inches long and one and a half inches wide, with fibers curling

Yucca baccata, banana yucca; young fruits. Note withered petals.

back along the margins. The leaf tips end in a stiff spine that can leave a nasty sting when it encounters your skin.

Depending on your imagination, banana yucca fruits resemble bananas, berries (hence the species name, *baccata*, meaning "adorned with berries"), or dates (*datil* means "date" in Spanish). The edible theme is no coincidence, since banana yucca fruits were widely used as a food source by many Native American peoples of the Southwest. They roasted the fruits and pounded the pulp into cakes, or patties. These little cakes were either eaten directly or dried for storage. The Acoma and Laguna cooked the fruits to obtain syrup, and the Ramah Navajo used them for preserves. The Apache ate flowers harvested before the summer rains, because afterward the flowers turned bitter.

Example of cordage. Artifact found in cave on Lower Pecos River.

In the medicine department, the Tewa ate the fruits for an easy childbirth. The Navajo made a tea of pounded leaves to stop vomiting or relieve heartburn.

Yucca roots contain sudsy saponins and were widely used as a source of soap. *Yucca baccata* roots,

Example of basket fragment. Artifact found in cave on Lower Pecos River.

however, were used by the Navajo only for washing ceremonial implements.

Banana yucca leaves were invaluable for their fibers. The leaves were prepared by boiling or beating them, soaking them for a few days, and then separating and cleaning the fibers. These fibers were used to make rope, fishnets, head straps, thongs for footwear, and baskets. The fibers were also woven into skirts and other clothing items and were used to tie the feathers of prayer plumes together.

The leaves themselves were split into strips for use in making baskets, mats, clothing, head pads, and sandals (see photos). The ends of the leaf strips were chewed to make paintbrushes, and thin strands of leaves became tools for decorating pottery.

Although banana yucca flowers are visited by various insects, including butterflies, they are pollinated only by the pronuba moth *Tegeticula baccatella*, which works at night and rests inside the flower during the day. The female gathers pollen from one flower, inserts it into the stigma of another flower, and deposits her eggs. The larvae feed on the developing seeds but destroy only between 10 and 15 percent of the seeds in the fruits.

Elk, deer, javelinas, and bighorn sheep browse on various parts of banana yucca. Small mammals, birds, and insects also consume parts of the plant, and small mammals and birds use it for cover. The ursine giant-skipper butterfly (*Megathymus ursus*) uses it as a host for its larvae, which feed on the tips of leaves and then move down to tunnel into the roots, where they pupate.

ASPARAGACEAE (Asparagus Family) (formerly in Agavaceae)

Yucca baileyi

NAVAJO YUCCA, BAILEY'S YUCCA, ALPINE YUCCA

Navajo yucca, found in northern New Mexico, grows at slightly higher elevations than other *Yucca* species, in desert grasslands, foothills, and canyons. Where its range intersects with that of another yucca, *Yucca glauca*, hybrid populations may be found. Navajo yucca resembles *Y. glauca*, but the flowers are held closer to the flower stalk, and the buds,

Yucca baileyi, Navajo yucca; young fruit.

Yucca baileyi, Navajo yucca; flower with pronuba moth.

Yucca baileyi, Navajo yucca; mature seedpods.

sepals, and pedicels are purple tinged. The narrow, flexible leaves have some fibers curling backward along the lower half of the leaf edges.

Native Americans pounded the roots with a stone to break up the fibers and release the sudsy saponins, and then added them to water for ceremonial bathing, cleaning baskets and feathers, washing wool for weaving, and washing their hair. They believed that yucca root shampoo strengthened and beautified the hair, while also preventing baldness and keeping away lice. Midwives used the soap from the roots to clean and lubricate their hands.

Like other yuccas, Navajo yucca is pollinated by its own personal pronuba moth, *Tegeticula altiplanella* (see photo). Among the other insect visitors is Strecker's giant-skipper butterfly (*Megathymus streckeri*).

ASPARAGACEAE (Asparagus Family) (formerly in Agavaceae)

Yucca elata

SOAPTREE YUCCA, PALMILLA

It's easy to feel elated when looking at the exalted heights of *Yucca elata*, crowned with spires of white flowers that sometimes reach to thirty feet.

Yucca elata, soaptree yucca; exposed root system at White Sands National Monument.

Soaptree yucca has thin, narrow leaves that can remain green for up to five years. A profusion of long, straight fibers hang from the lower third of the leaf margins.

Soaptree yucca is found in the lower half of New Mexico, growing in desert plains and grasslands up to 6,000 feet elevation, where it breaks the monotony of a flat skyline. It is a key feature of White Sands National Monument, with its long, thin trunk sometimes buried completely in the sand, or recumbent on its side, or with its roots partially exposed as the winds shift the sand around (see photo).

As with all yuccas, Native Americans used the leaves for cordage, such as the artifact shown in the photo. The Papago still use the thin leaves in their basketwork, drying and bleaching some of them in the sun to make the white coils and leaving other leaves unbleached for the green patterns.

Example of plaited rope. Artifact found in cave on Lower Pecos River.

Yucca elata is pollinated by pronuba moths, primarily the species *Tegeticula elatella*. Soaptree yucca also hosts larvae of the yucca giant-skipper butterfly (*Megathymus yuccae*).

Soaptree yucca's leaves are more tender than those of its cousins, and wildlife and livestock use them for browse during the winter and spring, when the young leaves are just emerging. Mule deer, pronghorn, jackrabbits, and rodents are especially fond of the flowers.

Soaptree's longevity in clonal groups makes it a long-term soil improver, with its leaf litter accumulating nearby and its aggressive root and rhizome system helping to stabilize desert soils. Plants are available commercially for use in landscaping.

ASPARAGACEAE (Asparagus Family) (formerly in Agavaceae)

Yucca faxoniana

EVE'S NEEDLE, SPANISH BAYONET, FAXON YUCCA

The species name of the very showy *Yucca faxoniana* honors Charles E. Faxon, who illustrated the nineteenth-century book *The Silva of North America*.

This plant is found in southeastern New Mexico in limestone soils from 4,000 to 5,000 feet elevation. The trunk is treelike, with a more squatty body than *Y. treculeana*, another arborescent species (see the next entry), and may have one or more branches. The inflorescence is a rounded panicle full of globose white flowers, about three-quarters of which extend beyond the top of the leaves. The long, stiff, pointed leaf

tips are nicely suited for Eve's sewing projects and have no fibers curling from the leaf margins.

The stately shape of Spanish bayonet is a key feature in xeriscapes in the Southwest.

ASPARAGACEAE (Asparagus Family) (formerly in Agavaceae)

Yucca treculeana (syn. *Yucca torreyi*)

TRECUL YUCCA, TORREY YUCCA, SPANISH DAGGER

"Trecul yucca" has been made synonymous with "Torrey yucca," by which name it has long been known. It was originally named for John Torrey, a nineteenth-century botanist who wrote extensively about the plant collections sent to him by explorers in the western United States.

Torrey yucca is found in the Chihuahuan Desert in the southern half of New Mexico, growing in the plains and on the tops and sides of mesas from 2,800 to 5,000 feet elevation. The thick stem, reaching to fifteen feet high, is often branched at the top, and when the plant is in bloom, the sight is spectacular, especially when the area is well populated with these statuesque yuccas.

The leaves are moderately rigid and somewhat concave, with numerous fibers curling from the leaf margins. The bell-shaped flowers are usually purple tinged and hang downward.

The red patterns in Apache basketwork are made with the roots of this yucca, and its leaves constitute the main portion of certain basket types.

Yucca treculeana, Torrey yucca; note numerous fibers curling from leaf margins.

Trecul yucca has a dedicated pronuba moth, *Tegeticula treculeanella*, to carry out the nuptials. Other insects visit the flowers, including the ursine giant-skipper butterfly (*Megathymus ursus*), whose caterpillar feeds on the ends of the leaves and moves downward to pupate in the roots.

ASTERACEAE (Sunflower Family)

Achillea millefolium

YARROW, PLUMAJILLO, MILFOIL

When the Greek warrior Achilles stormed Troy in the Trojan War, he used this plant to stanch the flow of blood from the combat wounds of his soldiers. Early on, it was dubbed *Achillea*, and it continued to be known as a soldier's herb from the time of the Roman Empire to the Crusades; it was also used extensively during the American Civil War. Its finely divided, feathery leaves (see photo) give it the species name *millefolium* (Latin for "thousand leaves") and the common name "plumajillo," which means "little feather."

Achillea millefolium, yarrow; finely divided, feathery leaves.

Yarrow grows in alkaline caliche to rich loam soils in a variety of vegetation communities from 5,500 to 11,000 feet elevation. The flowers bloom from April through July. It propagates to some extent by seed but most rapidly via rhizomes, making it a good rehabilitation tool for damaged land and erosion control.

Achillea is extremely morphologically variable, and taxonomists who are "splitters" recognize about sixty species. The "lumpers" treat it as a single species—a cosmopolitan, circum–Northern Hemisphere complex with diverse populations that hybridize freely.

Yarrow's bitter taste and pungent fragrance are clear indications that its best use is in the medicine chest and not in the kitchen. The leaves contain the astringent alkaloid achilleine, as well as menthol, camphor, and other compounds that are antimicrobial and analgesic. This chemical portfolio not only is useful against coughs and colds but also keeps bugs at bay, even killing mosquito larvae. Essential oil of *Achillea* is available commercially, and health stores carry yarrow tea.

Southwest tribes used yarrow as an analgesic for headache, toothache, and fever and as a poultice for burns and abscesses. Some inhaled the smoke from burning dried leaves as a decongestant and used its astringent properties to reduce hemorrhoids. They also washed saddle sores and cuts on their horses with yarrow tea.

The Navajo took yarrow as a ceremonial emetic, and the Zuñi applied juice extracted from the flowers and roots for cooling the throat and feet before fire-eating or fire-walking ceremonies. It is even said to be a love charm, especially if plucked by a maiden from the grave of a young man. That's enough to strike fear into the heart of the pursued!

An extensive number of cultivars of *Achillea* are available commercially in a range of bright and pastel colors, from yellow to red to pink, in addition to the traditional white, for use in xeriscapes and gardens. Yarrow is a good cut flower, providing a refreshing pungence, and is excellent in dried flower arrangements as well. It is a larval host for painted lady (*Vanessa cardui*) and chalcedon checkerspot (*Euphydryas chalcedona*) butterflies.

ASTERACEAE (Sunflower Family)

Aphanostephus skirrhobasis

ARKANSAS LAZY DAISY

Aphanostephus means "inconspicuous wreath" in Greek, referring to the crown-shaped pappus, or cluster of scaly hairs, atop the seeds. The base of the flower head, where the multitude of tiny flowers attach, becomes hard and thick with age, hence the species name, *skirrhobasis* (hard base).

The key claim to fame of Arkansas lazy daisy is its charming white flowers, which dance in the breeze like fairies. The reverse side of the

Aphanostephus skirrhobasis, Arkansas lazy daisy; field view.

white ray flowers (the ones that look like "real" petals) is pink, which is the part that shows early in the day. Lazy daisy is well named, for it sleeps in until midmorning, dressed in its pink pajamas, then wakes up and puts on its cute little white tutu until late afternoon, when it goes back to bed.

Arkansas lazy daisy grows in the Chihuahuan Desert of southeast New Mexico. It thrives in dry, rocky caliche soil almost as well as in well-watered soil. It is ideal in a garden, whether water availability is low or high, and has a long bloom period, from March to August, or even until frost if the plants are kept deadheaded.

ASTERACEAE (Sunflower Family)

Chaetopappa ericoides

ROSE-HEATH, SMALLFLOWER ASTER, BABY WHITE ASTER

Rose-heath's genus name, *Chaetopappa,* is based on the Greek words for "bristle" and "fluffy appendage," descriptors for the wild, hairy attachments (see photo) on the achenes, or small fruits. *Ericoides* tells us that it is "like *Erica,*" the heath plant.

Rose-heath is a doll-size plant with small leaves found throughout New Mexico. It thrives in well-drained, calcareous soil, as well as in moist conditions, in open sites in grasslands, chaparral, and pinyon-juniper communities up to 6,500 feet elevation. The vigorous underground rhizomes form low, unkempt mats decorated with dainty white flowers that bloom from March through November.

Native Americans of the Southwest used rose-heath to treat a variety of ailments. Swollen or inflamed body parts were poulticed with the whole plant. People with digestive difficulties bathed in a tea made from the whole plant or just the roots. For kidney and bladder disease, the Navajo drank a rose-heath tea mixed with sumac berries (see p. 262). They also sneezed out a cold with snuff made from the dried plant. The Hopi drank away a cold with a tea made from the roots. The Navajo chewed the leaves for a toothache and made a poultice of chewed leaves, along with tea, to treat snakebite.

Chaetopappa ericoides, rose-heath; note fluffy pappus on seed head.

The Hopi used a root tea to soothe distressed babies, and Zuñi mothers drank a warm tea to speed up a prolonged birth.

ASTERACEAE (Sunflower Family)

Erigeron spp.

FLEABANE

The fleabane group is very large and complex, making it difficult to distinguish one species from another. We have, therefore, chosen to discuss them generically, as *Erigeron* spp. Their overall appearance is similar, with multiple series of narrow ray petals, woolly leaves and stems, and an inescapable cuteness. The ray flowers are often white, sometimes lavender-blue, and occasionally totally pink.

The genus name, *Erigeron*, comes from the Greek *eri* (early), or possibly *erio* (woolly), and *geron* (old man), referring to its early flowering in the spring, or perhaps the hairy covering of its leaves and stems and the rapid maturation of the achenes, each with its attached, fluffy white pappus. "Young today and old tomorrow" could be its motto.

Erigeron is closely related to *Conyza* (see p. 142), with which it shares similar chemical constituents, and to *Aphanostephus* (see p. 282), with which it shares petal coloration (the underside of white petals is often a delicate pink) and a tendency to sleep in until almost noon, like night shift workers in their pink pajamas.

Erigeron spp., fleabane; showing pink pajamas.

Fleabane grows in a variety of soil types in desert to subalpine elevations. Several species are very attractive in home gardens or xeriscape plantings, and seed is commercially available for a number of varieties.

Erigeron's volatile oils have insect-repellent properties. Indians and early settlers burned the plant to drive away fleas (hence the name "fleabane"), gnats, and other insect invaders.

Fleabane was widely used as a household medicine among many Native Americans. Two common uses were as a diuretic and as a sudorific, to induce sweating. The Navajo used it in several chant ceremonies, as well as to treat coughs, fever, and a variety of other diseases.

Fleabane hosts a wide assortment of insects, including both larvae and adults of the northern crescent butterfly (*Phyciodes cocyta*).

ASTERACEAE (Sunflower Family)

Hymenopappus newberryi

WILD COSMOS, NEWBERRY'S WOOLLY-WHITE

Hymenopappus is a compound Greek word meaning "membrane bristle," referring to the very small scales at the apex of the achenes, or small fruits, of the members of this genus. John Newberry, whom Asa Gray honored with the species name, was a nineteenth-century professor who was the first geologist to explore and describe the Grand Canyon and, later, the upper Colorado and San Juan Rivers.

Hymenopappus newberryi can be found in all the mountain ranges in New Mexico. This perennial blooms from June to September. It grows in

clay and igneous soils in aspen, pine, and spruce communities at upper elevations. Plants reach knee high and have a basal rosette of hairy, finely divided leaves, with a few like-minded leaves on the equally hairy flowering stems. The flower heads have white ray flowers and yellow disk flowers that do have some resemblance to cosmos, from which the common name is taken.

The Isleta made a tea from this plant to treat stomachache.

ASTERACEAE (Sunflower Family)

Layia glandulosa

TIDYTIPS, WHITEDAISY TIDYTIPS

Sir William Hooker gave tidytips its genus name to honor George Lay, a nineteenth-century botanical explorer in the circum-Pacific countries. *Glandulosa* recognizes the many glands covering the phyllaries and the underside of the leaves.

The leaves and stems are hairy, and the plant often has a spicy scent. The white-rayed flower heads are borne on individual stems and look coquettish, or possibly snaggletoothed, with some of the ray petals dipping down out of line with the others.

Tidytips grows in dry sandy to rocky soil in deserts, grasslands, and woodlands up to 9,000 feet elevation, with flowers appearing from February to July. The plants usually reach less than two feet tall.

Birds use tidytips as a food source. Tidytips has excellent potential as a mounding plant or ground cover in waterwise landscapes.

ASTERACEAE (Sunflower Family)

Melampodium leucanthum

BLACKFOOT DAISY, PLAINS BLACKFOOT

The genus name of plains blackfoot has two possible derivations. The first is strictly botanical: *melampodium* (black foot) refers to the dark achenes that form at the base of the ray flowers. (The disk flowers are usually male and do not set seed.) These small, dry fruits do not have the usual hairy pappus, but instead have one of the inner rings of phyllary bracts attached, forming a "hood" above the fruit. This turns brownish black with maturity and looks somewhat like a pointy-toed black slipper (the "black foot"). (From another angle, one can easily imagine a toilet seat on a stick.) Another source suggests a second derivation, from

Melampodium leucanthum, blackfoot daisy; mature seed head.

Melampus, a mortal famed in Greek mythology for his ability to understand the speech of animals, as well as for his herbal medicine knowledge. His name also means "black foot." There's no controversy about the species name, *leucanthum*, which is Greek for "white flowers."

There is also no doubt about the extraordinary gift that blackfoot daisy has for withstanding heat and drought while continuing to put forth an abundance of flowers. This perennial with a deep taproot grows in dry limestone or sandy soil in disturbed land and in grassland, chaparral, and pinyon-juniper communities up to about 5,000 feet elevation. Its growth habit is mounding, and fragrant white-rayed flowers cover it from spring to fall. It self-seeds readily and grows quickly to about ten inches tall.

Blackfoot daisy is increasingly popular in natural and waterwise landscapes. Since it is a low plant, plan for it to be at the front of the garden or in a border. Be aware that mollycoddling blackfoot daisy with generous water and fertilizer will shorten its life span. For garden enthusiasts, it is important to note that blackfoot daisy is said to be very deer resistant.

It is also popular with insects, especially bees and butterflies, which enjoy its nectar. Small birds and mammals eat the seeds.

ASTERACEAE (Sunflower Family)

Parthenium incanum

MARIOLA

Parthenos is the Greek word for "virgin," which may, or may not, be related to why Theophrastus gave the name *Parthenium* to Asteraceae plants with white ray flowers. While the significance of the genus name is obscure, the connotation of the species name is clear: *incanum* means "gray," in reference to the fine hairs that give the leaves a silvery look.

Mariola is a small shrub that resides in a tough environment of rocky desert grasslands and chaparral, usually in alkaline or calcareous soil, from 2,800 to 6,500 feet elevation. It is one of the major plants of the Chihuahuan Desert and can be dominant in some areas. The flowers appear from May through November in small, white flower heads that resemble little cauliflowers.

Mariola is closely related to, and hybridizes with, *Parthenium argentatum*, or guayule, a minor commercial source of natural rubber. Guayule contains 10 to 12 percent of its dry weight as latex, compared to only 1 to 2 percent in mariola. The latex is produced in the bark along with sap resins containing terpenes and quinine-like alkaloids.

The bitter and aromatic stems and leaves were a source of medicines used by Southwestern tribes. The Apache boiled the fresh leaves for a beverage similar to coffee.

Mariola's shape and silver foliage are attractive in a xeriscape or natural landscape. It is also used for erosion control on roadsides and road cuts.

ASTERACEAE (Sunflower Family)

Townsendia incana

EASTER DAISY, HOARY TOWNSEND DAISY

These petite plants are totally adorable, with white blooms performing in concert with mature flower heads that spill out many white-fluffed seeds. Most species play only one instrument at a time—first the flowers, then the fruits. The simultaneous performance of Easter daisy makes it look a bit disheveled, but that doesn't detract from its spot atop a list of desirable landscaping plants.

The genus name honors the Pennsylvanian David Townsend, who lived in the early 1800s and was by profession a banker and by heart a botanist. In his memorial, it was said of him that the same traits of detailed examination that made him a successful banker served him equally well in investigating the intricacies of the botanical world. *Incana* means "gray," describing the hairy white stems and leaves. The leaves are about the length of a cat's tongue but folded in the middle, and from the side they look like geese's heads on long necks.

Easter daisy grows in the northwest counties of New Mexico, in sandy or rocky soil in desert to pinyon-juniper communities from 3,500 to 7,000 feet elevation. Its flowering season is April to July.

The Hopi and Navajo believe this plant can help a woman conceive (particularly a male child) and ease childbirth. The Navajo use the plant in the Water Way Ceremony and the Night Chant.

BORAGINACEAE (Forget-me-not Family)

Cryptantha flavoculata

YELLOW-EYE CRYPTANTHA, DESERT POPCORN, ROUGHSEED CRYPTANTHA

Cryptantha, Greek for "hidden flower," aptly describes the tiny flowers in this genus. The "yellow eyes" (*flavoculata*) are tiny appendages on the white petals, sitting like a little golden crown above the floral tube and making the flower resemble a piece of popcorn.

The flowers in the forget-me-not family are arranged in a coil called a cyme, resembling the scroll of a cello. In many species, this one included, the flowers are crowded among so many white hairs that the coil pattern is difficult to discern.

Desert popcorn is a perennial that grows in western New Mexico in desert, sagebrush, and pinyon-juniper communities. It blooms in April and May.

The Hopi used this plant to treat cancer of the throat. The Navajo used it in powdered form to soothe sore eyes, for purification of women after childbirth, and to treat inflammation of the intestines.

This species is one of the more attractive members of *Cryptantha* and makes a colorful addition to a waterwise landscape. *Cryptantha* attracts insect visitors such as the pterophorid plumed moth shown in the photo.

BORAGINACEAE (Forget-me-not Family)

Heliotropium convolvulaceum

BINDWEED HELIOTROPE, PHLOX HELIOTROPE, FRAGRANT HELIOTROPE

This largest of all heliotropes is covered in the summer and early fall with snowy white blooms that are arranged in small coils. In the center of the white corolla is a tiny yellow "eye," reminiscent of the "eye" in a phlox flower—the reason "phlox heliotrope" is included in this species' repertoire of common names. The pleasant-scented flowers open in the afternoon.

Fragrant heliotrope's genus name means "turning" (*trope*) "toward the sun" (*helios*). The species name, *convolvulaceum*, recognizes the

Heliotropium convolvulaceum, phlox heliotrope; close-up of maturing fruit.

similarity of the flowers to those of plants in the morning-glory family (Convolvulaceae). It grows in dunes and sandy expanses in desert and grassland communities and is found on the east and west sides of New Mexico's central mountains from 4,000 to 6,000 feet elevation.

Monarch (*Danaus plexippus*), queen (*Danaus gilippus*), and soldier (*Danaus eresimus*) butterflies use fragrant heliotrope as a nectar source.

BRASSICACEAE (Mustard Family)

Dimorphocarpa wislizenii

SPECTACLE POD, TOURISTPLANT

The genus name tells us that some species in the genus have "two forms" (*dimorphus*) of "fruit" (*carpus*), although that is not the case for spectacle pod. The species name honors Friedrich Wislizenus, a nineteenth-century German physician and naturalist who immigrated to the United States, where he participated in plant-collecting expeditions until his middle age, when he settled down to the life of a physician.

The most captivating aspect of this plant is the seedpods, which look like old-fashioned opera glasses or Mardi Gras masks. The white flowers are clustered at the top of the eight- to twenty-inch-tall stem, and the seedpods develop below them, until the stem is full of pairs of green eyes, as shown in the photo. The plant begins to flower in early spring

Dimorphocarpa wislizenii, spectacle pod; note two-chambered seedpods.

and continues through October. It grows throughout New Mexico along sandy roadsides and in other sandy locations from 3,500 to 6,500 feet elevation.

Spectacle pod was used by the Acoma and Laguna to treat congested noses: they crushed the seeds and leaves and inhaled the powder to bring about sneezing. The Navajo used it for soothing and shrinking hemorrhoids, and their children were encouraged to chew this plant to strengthen their teeth. The Ramah Navajo used a spectacle pod infusion as a lotion for itchy skin, including chicken pox, measles, insect bites, centipede stings, and cold sores. The Zuñi made a poultice to reduce swelling and treat sore throat, and they drank spectacle pod tea as an emetic to relieve stomachache.

BRASSICACEAE (Mustard Family)

Lepidium alyssoides (syn. *Lepidium montanum*)

MOUNTAIN PEPPERWEED, PEPPERGRASS

Lepidium means "a little scale," referring to the flat shape of the seedpods, as shown in the photo. The species name, *alyssoides*, tells us that the tiny white flowers are arranged in long racemes that resemble those of

Lepidium alyssoides, mountain pepperweed; close-up of inflorescence.

Alyssum, another genus in the mustard family. The peppery taste of the leaves and seeds gives it the common name "pepperweed."

Pepperweed grows throughout most of New Mexico, in a variety of soil types in grassland, sagebrush, and pinyon-juniper communities from 3,500 to 8,500 feet elevation. It flowers from May to July.

Although pepperweed is not part of the modern Native American diet, its earlier importance as a food source is evidenced by archaeological findings of *Lepidium* seeds in coprolites. Pepperweed seeds, which taste a bit like a hot, spicy version of broccoli, are quite tasty in a salad. The Kayenta Navajo took pepperweed for digestive disorders, as well as for heart palpitations and dizziness. The Isleta chewed pepperweed leaves to soothe a headache. The Kayenta Navajo rubbed the plant on a baby's face to make him or her sleepy.

Lepidium alyssoides, mountain pepperweed; scalelike seedpods.

Pepperweed is a food source for both adults and larvae of the desert marble butterfly (*Euchloe lotta*) and larvae of the great southern white butterfly (*Ascia monuste*).

BRASSICACEAE (Mustard Family)

Nasturtium officinale

WATERCRESS

If *Nasturtium* sounds a bit nasty to you, it also did to the person who named it. The genus name comes from the Latin *nasus* (nose) and *tortus* (twisted or distorted)—that is, the pungent taste of the plant twists one's nose! The species name, *officinale*, is derived from *officina*, Latin for "workshop" or "laboratory," in reference to a herbal pharmacy, indicating that the plant was sold as an herbal medicine. Indeed, the list of illnesses for which *Nasturtium* was traditionally used in its European home is lengthy: anemia, boils, kidney and liver disease, tuberculosis, tumors, and warts. It also served as a diuretic, a purgative, and a stimulant.

It is now cultivated commercially because of its popularity as a salad ingredient with a slightly peppery flavor. It is high in vitamin C, which led to its historic use as a treatment for scurvy.

Native Americans used watercress both as a vegetable, eaten either raw or cooked, and as a medicine. The Costanoan of the West Coast used watercress as a liver and kidney remedy, and the Okanagan-Colville made a poultice of the plant to treat headache. The Mohegan made a poultice to place on the jaw of someone suffering from tooth pain. Watercress is said to be effective against rheumatism, sore throat, and stomachache.

In spite of its many valuable uses, watercress, like dandelion, is much maligned as a weedy and invasive pest. This aquatic perennial is common throughout New Mexico in streams, marshes, ditches, and other wetlands from 5,000 to 8,000 feet elevation. Its white flowers appear from April to October on three-foot-long floating stems that root at each node and form new leaves.

BRASSICACEAE (Mustard Family)

Nerisyrenia camporum (syn. *Greggia camporum*)

BICOLOR FANMUSTARD, VELVETY GREGGIA, VELVETY NERISYRENIA

Although the origin of the name *Nerisyrenia* is obscure, the species name, *camporum*, tells us that this is a plant of the "plains" or "fields" (*campus*). The common name "bicolor fanmustard" alludes to the two colors of the petals.

Fanmustard flowers bloom from February through October and are white or very light lavender, with the petals becoming pink as they age. The arrangement of the four petals resembles a crucifix, on which the former family name, Cruciferae, was based. A close inspection of the pale bloom reveals quaint, curly yellow stamens that give a whimsical appearance to the flowers. The seeds are coated with mucilage, which becomes thick and sticky when wet.

Bicolor fanmustard is found in southern New Mexico, where it grows as a two-foot-tall perennial in dry clay, sandy, or alkaline soil in desert and grassland communities from 3,500 to 5,000 feet elevation. The plant surfaces are covered with white, velvety, star-shaped hairs.

Also pictured here is a white-flowered relative, *Nerisyrenia linearifolia*, which grows in gypsum soils in southern New Mexico along with bicolor fanmustard.

Nerisyrenia linearifolia, White Sands fanmustard; close-up of flower and seedpods.

BRASSICACEAE (Mustard Family)

Thelypodium wrightii

WRIGHT'S THELYPODIUM, WRIGHT'S THELYPODY

Thelypodium means "female foot," said to refer to the long stipe, or stalk, above the female part of the flower. The species name honors Charles Wright, a nineteenth-century botanical collaborator of the famed Asa Gray and John Torrey, who collected extensively in Texas and Mexico, as well as in the Pacific Northwest.

Thelypody is five feet tall and grows in the northern third and western half of New Mexico in pinyon-juniper and oak woodlands from 5,000 to 6,000 feet elevation. The clusters of white flowers produce very long seedpods, called siliques, that droop and sway gracefully, as shown in this photo from Bandelier National Monument.

Ancestral Puebloans consumed *Thelypodium wrightii* as a vegetable, boiling and eating it fresh and also storing some for later use by rolling it into balls that were then dried. Zuñi women crushed the seeds into a meal to make mush.

The Navajo used thelypody poultices to reduce swelling, and the Kayenta Navajo made an eye medicine from the plant. The Kayenta also tied thelypody to cradles to help babies sleep.

CACTACEAE (Cactus Family)

Cylindropuntia leptocaulis (syn. *Opuntia leptocaulis*)

CHRISTMAS CACTUS, TASAJILLO, PENCIL JOINT CHOLLA

The genus name of Christmas cactus combines the description of its "cylindrical" stems with an early name given to members of the cactus family, *Opuntia*. The "slender stems," indicated by the species name, *leptocaulis*, are a mere one-eighth to one-quarter inch thick and have solitary spines. The flowers bloom from March through October and are the least notable feature of this plant—a half inch of yellowish-white color that appears late in the day.

The plant grows in desert, chaparral, and pinyon-juniper communities from 3,000 to 5,500 feet elevation, its pencil-like stems enmeshed with those of more robust plants such as creosote bush (*Larrea tridentata*; see p. 121). The small but eye-catching red fruits shine out from this tangled embrace, earning it the name "Christmas cactus" because the fruits overwinter on the plant, looking like Christmas tree ornaments.

The Pima eat the small red fruits (with the prickles carefully removed), which are high in vitamin C, as are all cactus fruits.

Like other cacti, Christmas cactus can play a colorful role in a native landscape. Cactus wrens use it for nesting, and deer, turkey, quail and other small birds, and small mammals use it for a food source as well as for cover.

Christmas cactus has a special relationship with creosote bush, which colonizes a site and modifies the microclimate, making it favorable for Christmas cactus establishment. Over time, Christmas cactus suffers the predation of burrowing mammals and soil erosion, paving the way for

Cylindropuntia leptocaulis, Christmas cactus; showy red fruits.

creosote bush to recolonize the openings.

Two different red dyes are associated with *Opuntia* and *Cylindropuntia* species. One, a red to blue betalain dye, is extracted from cactus fruits, as well as from members of the amaranth family, and is approved by the FDA as a food colorant. The other, a bright red dye, comes from the cochineal scale insect *Dactylopius coccus,* which actually grows on *Opuntia* and *Cylindropuntia* species. In this fascinating host-guest relationship, the insects conceal themselves under white, frothy fuzz that looks like bird droppings on the surface of the cactus (as shown in the photo). When crushed, they release a pigment that is as bright as blood. The Aztec and their neighbors used this pigment to dye fabric prior to the Spaniards' arrival in the New World. The Spaniards were enchanted by this rich color and created a market in Europe for the cochineal dye. The export value of the dye was second only to that of gold, and Spaniards had a monopoly on the trade for the better part of two centuries, until the 1770s. Today the cochineal scale insect is still cultivated on prickly pear, and the dye is widely used as a natural colorant for food, beverages, textiles, and cosmetics throughout the world.

For information about another *Cylindropuntia* species, see p. 228.

Cylindropuntia leptocaulis, Christmas cactus; with white excrescence from cochineal scale insects.

CLEOMACEAE (Cleome Family) (formerly in Capparidaceae)

Polanisia dodecandra

CLAMMYWEED, REDWHISKER CLAMMYWEED

Redwhisker clammyweed is a lovely plant with long red stamens and decorative seedpods. *Polanisia* has "many" (*poly*) "unequal" (*anisos*) stamens, some of which are very long but all of which are well exserted beyond the petals. The species name, *dodecandra*, specifies twelve stamens (literally, "twelve males"), but in fact the number ranges from ten to twenty.

The entire plant is covered with short, sticky, stalked, gland-tipped hairs that give it a clammy feel. Like its cousin *Cleome serrulata* (see p. 234), it has palmate leaves with three leaflets (see photo). The white flowers are in racemes, or long clusters, that bloom from May through

Polanisia dodecandra, clammyweed; note palmately compound leaves.

Polanisia dodecandra, clammyweed; note dehiscing (splitting) seedpods.

October. Clammyweed grows in sandy areas in various plant communities up to the pinyon-juniper zone.

Clammyweed was used in much the same ways as *C. serrulata*. The Puebloans of New Mexico boiled the young plants as fresh greens or rolled them into balls to dry for winter use. Both fresh and dried clammyweed were used in stews with bits of meat, tallow, and wild onions. The Isleta rolled clammyweed inside corn husks to make cigarettes, which they smoked during ceremonies. The roots and flowers were chewed and rubbed on the body during the Cactus Fraternity Ceremony.

EUPHORBIACEAE (Spurge Family)

Euphorbia albomarginata (syn. *Chamaesyce albomarginata*)

RATTLESNAKE WEED, WHITEMARGIN SANDMAT, SPURGE

Euphorbia is named for Euphorbus, the Greek physician who served King Juba II of Mauritania two thousand years ago. King Juba was interested in botany and was especially fond of a low-growing spurge whose latex sap he used as a medicine. He honored his physician by naming the plant *Euphorbia*. The word itself means "good food," evidently a play on words, because Euphorbus was as plump as the plant's fruits. The species name, *albomarginata* (white border), refers to the narrow white margin of the leaf, although the white floral bracts would be a more obvious choice for the epithet.

Rattlesnake weed forms dense mats in disturbed soil throughout New Mexico at less than 7,000 feet elevation. Decorating the surface are

Euphorbia albomarginata, rattlesnake weed; whole plant.

the flowers, which are little white "cups" with pink "glands" around the edge. The plants are monoecious, the female flowers having a large ovary hanging out, and the males having obvious stamens. They bloom from April through November.

Euphorbia species produce a milky sap that contains many pharmacologically active chemicals that give the plant medicinal value. The sap has long been a traditional remedy for corns, waxy growths, warts, asthma, and sinus infections, along with other mucous membrane inflammations. Many members of the *Euphorbia* genus contain ingenol angelate compounds, which are being investigated for their potential to promote wound healing and to treat HIV and cancer, particularly leukemia, melanoma, prostate cancer, and cervical cancer brought about by human papillomavirus (HPV). An extract of the plant's sap is used to treat genital warts caused by HPV.

Archaeologists have found stored seeds of *Euphorbia* species at Salmon Ruins in the Four Corners area. The Navajo used a cold tea from rattlesnake weed for stomachache. They also used the plant as a hemostatic poultice. Several tribes either made a poultice of the crushed whole plant or ground the leaves and flowers to make a salve to treat rattlesnake bite. Other uses include rubbing the leaves on the breasts of nursing mothers and eating the leaves and roots to stimulate milk production. A tea made from the plant was used as a cleanser for skin sores.

Rattlesnake weed is an important food source for the desert tortoise (*Gopherus agassizii*).

FABACEAE (Bean Family)

Astragalus bisulcatus

TWO-GROOVED MILKVETCH

Students of anatomy will be familiar with the little tarsal bones in the foot, among which is the astragalus bone, from which this genus derives its name. The bone is somewhat kidney-shaped, like the seeds of the genus *Astragalus*. Roman soldiers used the bones as dice, which when shaken made a sound similar to the rattle of the seeds in the pods of a number of *Astragalus* species. The species name, *bisulcatus*, refers to the

Astragalus bisulcatus, two-grooved milkvetch; field view.

Astragalus bisulcatus, two-grooved milkvetch; note ridge and groove on developing seedpods.

"two furrows" on either side of a ridge along the top of the seedpod, as shown in the photo.

Two-grooved milkvetch is a perennial found in the northern half of New Mexico at elevations up to 8,000 feet. It can become invasive in poorly managed rangeland because livestock avoid it unless adequate forage is lacking.

There are about 1,500 species of *Astragalus* around the world, of which about 25 percent are found in the western United States and cause significant economic loss to livestock producers because of their potent toxins. About forty U.S. species cause severe damage to the nervous system because of their production of swainsonine alkaloids or their ability to intensively accumulate the heavy metal selenium.

Swainsonine allows the buildup of dangerous levels of the sugar mannose in nerve cells by blocking the action of the enzyme mannosidase. This latter action is similar to that of the recessively inherited disease mannosidosis, which results in paralysis and death at an early age. Animals that consume *Astragalus* for two weeks to a month will exhibit symptoms of "locoism," or swainsonine toxicity, which include extreme excitability, acting loco (crazy), and loss of muscle coordination. In spite of its effects, horses and other animals can become addicted to *Astragalus*.

Astragalus lonchocarpus, rushy milkvetch; whole plant.

Astragalus bisulcatus is also toxic due to selenium accumulation, which causes poisoning related to selenium's structural similarity to sulfur. Called "blind staggers" or "alkali disease," selenium poisoning leaves the animal unable to eat or control motor movements. Two-grooved milkvetch is an indicator plant for selenium to help identify soils rich in this mineral. The plant gives off the unpleasant smell associated with high concentrations of selenium. Hyperaccumulator plants such as two-grooved milkvetch tolerate selenium by creating a sulfur-selenium compound that is difficult to incorporate into proteins.

Two-grooved milkvetch is host to the caterpillars of Queen Alexandra's sulphur (*Colias alexandra*) and arrowhead blue (*Glaucopsyche piasus*) butterflies.

Another white-flowered *Astragalus* that produces swainsonine is *A. lonchocarpus,* pictured here.

For information about other *Astragalus* species, as well as the related genus *Oxytropis,* see pp. 193 and 311.

FABACEAE (Bean Family)

Dalea candida

WHITE PRAIRIE CLOVER

The genus name honors Samuel Dale, a British physician and botanist of the seventeenth and eighteenth centuries. The flowers are "white" (*candida*).

White prairie clover is a subshrub with two-foot-long stems rising from the crown at ground level. The plant grows in grasslands as well as disturbed places. The shape closely resembles that of purple prairie clover (*Dalea purpurea*; see p. 195), but white prairie clover looks more "jumbled" because the flowering stems branch near the top, creating a tangled effect. The flowers bloom from May through September, and then form tiny pods containing just one seed.

Dalea candida, white prairie clover; close-up of flowers and seedpods.

Navajo children and shepherds, as well as the Tewa of the San Ildefonso and Santa Clara Pueblos, learned that the long prairie clover taproot, which can extend as much as six feet deep, makes a nutritious, sweet snack eaten raw. The Acoma and Laguna dried the root and ground it into a meal used in baking.

White prairie clover is one of the Life Medicines of the Navajo. In the past, they used it to treat a number of health issues, including diarrhea, toothache, stomachache, and fever. It was also used to poultice arrow wounds. To prevent hair loss, the Acoma and Laguna made an infusion of the roots to wash their hair. They also used the many-branched plant as a light broom.

The plant is high in protein and highly palatable and will therefore decrease in overgrazed pastures. White prairie clover is host to larvae of the southern dogface butterfly (*Zerene cesonia*). It also attracts many other butterflies, as well as birds.

For information about other *Dalea* species, see pp. 195 and 242.

FABACEAE (Bean Family)

Desmanthus illinoensis

ILLINOIS BUNDLE FLOWER, PRAIRIE BUNDLE FLOWER, PRAIRIE MIMOSA

Illinois bundle flower is one of the most important native prairie legumes in the United States. Although its name features the state of Illinois, its range includes central and southern New Mexico. Its soil preferences are eclectic, from clay to sandy to caliche. It grows in water-blessed plant communities from about 3,000 to 6,500 feet elevation.

This three-foot-tall perennial blooms from May through September, its white flower clusters, punctuated with yellow stamens, decorating

Desmanthus illinoensis, Illinois bundle flower; dehisced seedpods showing seeds.

the stems. The stamens steal the show, with the actual petals and sepals relegated to the background. The flowers morph into a tight cluster of flat, curved seedpods, which are ornamental in their own right (see photo). Illinois bundle flower is one of the members of the mimosa tribe of the bean family with sensitive leaves that very rapidly fold together when touched. The bipinnately compound leaves give the plant an airy, lacy look.

The Pawnee boiled bundle flower leaves and used the extract to soothe itchy skin. The Paiute had an unusual (and horrific!) method of treating the eye disease known as trachoma: they placed five bundle flower seeds under the eyelid overnight, and then washed the eye in the morning. The Hopi also placed seeds in the eye to cure conjunctivitis.

Native American children used the hard, dried clusters of seedpods with the loose seeds inside as rattles to mimic ceremonial dancers.

Livestock, as well as deer and pronghorn, find this high-protein legume very palatable, although its tannin content reduces digestibility. The presence of Illinois bundle flower in a pasture is an indicator of good range management. The seedpods are filled with tick-shaped brown seeds that bruchid weevils adore and that provide food for quail and other birds, as well as rodents. A number of butterfly species feed on the flowers' nectar.

Illinois bundle flower is recommended for seeding in range revegetation projects. USDA programs in several states have released improved cultivars, and locally adapted seed is commercially available.

FABACEAE (Bean Family)

Oxytropis sericea

SILKY LOCOWEED, SILVERLEAF LOCO, WHITE CRAZYWEED

To explain the meaning of the genus name, a word about bean flower anatomy is in order. *Oxytropis* is part of the Papilionoideae tribe of the bean family, which we normally associate with peas and which has unique petal structures. The big, "standing-up" petal is called the banner, which appears to have two wings on each side. Between the wings are two petals fused into the shape of a boat's keel. It is this structure in *Oxytropis* (meaning "pointed keel") that looks like a long Pinocchio nose. The leaves and stems are densely covered in "silky" (*sericea*) hairs, giving the plant a silvery look.

Oxytropis sericea, silverleaf loco; pubescent (hairy) seedpods with remnant of style.

Silky locoweed is a perennial that grows to about fifteen inches tall throughout New Mexico in open areas from 6,500 to 9,000 feet elevation. It blooms from June through September.

Of all the plants poisonous to domestic animals, *Oxytropis,* along with *Astragalus,* is the most economically destructive to the livestock industry in the western United States. These plants cause extensive reproductive system failures, with cattle and particularly sheep aborting after eating *Oxytropis*. Most damaging of all is the permanent neurological damage to cattle, sheep, and especially horses. The toxic agent is swainsonine, an indolizidine alkaloid that blocks the enzyme mannosidase, which breaks down the metabolic sugar mannose in cells. As the mannose concentration rises, the damage to nerve cells increases, evidenced by a constellation of neurological symptoms, and becomes permanent. Affected horses are unsafe to ride because they act loco (crazy): they startle easily, and they run into objects because of damage to the optic nerve. Strangely, livestock become addicted to this agent of destruction and will seek out the plants even when good forage is available.

The Blackfoot of the northern Great Plains were among the few tribes that used silky locoweed. They made an infusion from the leaves to treat skin sores and ear troubles.

Silky locoweed fixes atmospheric nitrogen through the *Rhizobium* bacteria associations in its root zone. It serves as a larval host for the western tailed-blue butterfly (*Cupido amyntula*).

Oxytropis sericea has a purple-flowered cousin, *Oxytropis lambertii* (see photo), that has similar devastating effects on livestock.

For information about *Astragalus* species, which share some characteristics of *Oxytropis*, see pp. 193 and 305.

Oxytropis lambertii, purple loco.

GERANIACEAE (Geranium Family)

Geranium richardsonii

RICHARDSON'S GERANIUM, CRANESBILL

The species name honors Sir John Richardson, a nineteenth-century Scottish naval surgeon and naturalist who led expeditions to the Canadian Arctic, making extensive surveys of the coastline and writing prolifically about its flora and fauna.

The common name "cranesbill" calls attention to the unique seedpods, which bear an obvious resemblance to the head of the long-beaked bird. The "beak" is actually the expanded tubular style, which is sensitive to changes in moisture and catapults the seeds more than three feet from the mother. Once each seed lands, the coiled attachment twists in response to humidity, digging the seed into the soil.

This twenty-inch-tall rhizomatous perennial grows in rich, moist soil in ponderosa pine to spruce-fir zones from 8,000 to 11,800 feet elevation. The five white petals are delicately marked with lavender veins and have tangled hairs at the base. The bloom period is April to October. Dense, aromatic glandular hairs cover the flower parts and stems.

Geranium richardsonii, Richardson's geranium; large, palmately divided leaves.

Geranium's active ingredients are tannic and gallic acids, whose astringent properties led to many

Geranium caespitosum, purple geranium; meadow view.

medicinal applications for the plant. It is honored by the Ramah Navajo as one of their Life Medicines. The Cheyenne prized this plant for its ability to be made into a snuff to stanch nosebleeds. It was most often used for diarrhea, chronic dysentery, and cholera.

Bees are the primary pollinators, and flower beetles also participate in that ritual. Like its relative *Erodium cicutarium* (see p. 199), Richardson's geranium is palatable and nutritious forage for livestock and wildlife. It is useful for erosion control and revegetation projects.

Also shown here is a relative, *Geranium caespitosum*, a perennial living in moist, partially shady locations at higher elevations, from 6,000 to 9,000 feet. The fragrant leaves are foraged by sheep.

Geranium caespitosum, purple geranium; developing seedpods with long style column typical of geranium family.

GROSSULARIACEAE (Currant Family)

Ribes cereum

WAX CURRANT, WHISKY CURRANT

When Linnaeus named this plant in the 1700s, he apparently drew on a Latin word derived from the Arabic *ribas*, which means "sour" or "acidic," referring to the acidic taste of the juice of the berries, or "rhubarb," again in reference to the acidic taste of the fruit. The species name, *cereum*, means "waxy," in recognition of the coating of the red fruits.

Ribes cereum, wax currant; note waxy red berries and palmately veined leaves.

This perennial shrub grows to about five feet tall in open areas of forested regions of New Mexico from 5,000 to almost 13,000 feet elevation. The flowers bloom from April to June, and the waxy, bright red berries appear by August (see photo).

The Navajo, Acoma, and other tribes value wax currant berries as food, whether fresh, cooked, or preserved. Some Native Americans included the berries in pemmican. The Hopi used wax currant as a treatment for stomachache, and the Kayenta Navajo used it to clear out their digestive tract prior to certain ceremonies. They also used it as a poultice for sores.

The strong, straight branches were used by the Navajo to make arrow shafts and by the Tewa to make bows. Wax currant plants were a key element in the Navajo agricultural calendar: when the buds began to turn green, it was time to plow, and when the leaves appeared, it was time to plant corn.

Wax currant is an attractive ornamental component of natural landscapes. A number of birds eat the fruits. Wax currant is host to the larvae of the hoary comma (*Polygonia gracilis*) and tailed copper (*Lycaena arota*) butterflies. It provides some forage for deer and gophers, as well as for livestock, which will eat it when nothing better is on offer.

White pine blister rust (*Cronartium ribicola*), a fungal disease that attacks ecologically important high-elevation white pines, uses wax currant as the secondary host in its life cycle. Despite many attempts to eradicate wax currant in an effort to control the rust disease, both wax currant and white pine blister rust persist.

HYDRANGEACEAE (Hydrangea Family)

Fendlera rupicola

FENDLERBUSH, CLIFF FENDLERBUSH

The 1800s saw an extraordinary interest in collecting and documenting the plant life of western North America. One of the explorers in this golden age of botanical exploration was Augustus Fendler, a German immigrant whom Asa Gray commissioned as a plant collector to travel to Santa Fe in the 1840s. During his year there, he collected more than one thousand specimens within walking distance of the town. This and many other species were named in his honor.

The species name, *rupicola* (cliff dweller), as well as one of the common names, tells us that this plant grows on cliffs, which are plentiful in the Organ Mountains, where one of the photos was taken. Cliff fendler-

bush inhabits communities of desert shrubland to pinyon-juniper from 3,000 to 7,000 feet elevation. This shrub grows four to six feet tall and has showy white masses of flowers from March through June.

The Navajo drank an infusion made from the inner bark when they "swallowed a red ant," which was a way of saying they had an inner burning or stinging caused by bladder or kidney problems. They also used it to kill hair lice, which, according to archaeological findings, have been plaguing people as far back as three thousand years ago.

The Navajo used fendlerbush as an emetic in many rituals, including the Plume Way, Mountain Chant, Male Shooting Way, and Night Way Ceremonies. A notched stick of fendlerbush wood was rubbed with a smooth stick to accompany the Mountain Chant Ceremony.

Fendlera rupicola, fendlerbush; field view in Organ Mountains.

Fendlera rupicola, fendlerbush; old dehisced seedpods.

The strong yet flexible branches are straight and harden when heated, making them an excellent material for arrow shafts, weaving forks, planting sticks, looms, and knitting needles.

Both wildlife and livestock enjoy fendlerbush, which is very palatable to deer, bighorn sheep, cattle, and goats and may be overgrazed because of its desirability. It is also valuable as an ornamental plant in xeriscapes.

LAMIACEAE (Mint Family)

Marrubium vulgare

HOREHOUND

The genus name, *Marrubium,* is said to come from the name of a village where the city of Rome was eventually built. Others say that it is derived from the Hebrew word *marrob,* meaning the "bitter juice" of the herbs Jews consumed for the Passover feast. The species name, *vulgare,* means "ordinary." The "hore" in "horehound" refers to the hoary covering of fine white hairs on the leaves and stems.

Horehound is found throughout New Mexico, growing happily in dry calcareous soil below 7,000 feet elevation. The flowers are arranged in tight clusters along the stems like those of *Monarda* (see p. 202) and bloom throughout the summer, from June through September. The mature seed heads are like Velcro in their stickiness.

Horehound has a long history of use in the Old World as a flavoring for candy and ale. In modern times, horehound is grown commercially in France, and it has FDA approval as a "generally recognized as safe" (GRAS) natural flavoring in food.

In some regions of the Old World, horehound was a panacea—a cure for whatever was ailing a person, from skin sores, rheumatism, and malaria to coughs, colds, and hemorrhoids. When horehound was introduced to the New World, Native Americans found it to be effective in much the same ways. The Navajo brewed a horehound tea to treat colds, influenza, sore throat, stomachache, lameness, and general muscle pain. The Paiute used horehound branches to whip aching body parts, thus increasing circulation.

The Navajo still use horehound before and after childbirth, and it is one of the aromatic plants used in the Shooting Way and Wind Way Ceremonies. These plants are first put on the patient's body and then made into a tea for the patient to drink.

Horehound is an easy-to-grow garden or landscaping plant, and in fact it may prove too successful and become weedy. Bees and other insects are attracted to its nectar.

LAMIACEAE (Mint Family)

Teucrium laciniatum

CUT-LEAF GERMANDER

The genus name, *Teucrium*, honors Teucer, the king of Salamis in Cyprus. Pliny, the great first-century botanist, credited him with the discovery

of this plant as a medicinal herb. The leaves are "deeply cut" (*laciniatum*), giving the plant a frilly look. "Germander" was the name given by apothecaries (pharmacists) of the Middle Ages to plants useful for healing.

Cut-leaf germander has big flowers relative to the diminutive size of the plant, which is usually only six to eight inches tall. The corolla has five lobes, one of which is long and tonguelike. The plant is almost smooth, but there is a dense tuft of hair in the throat of the bloom. The sweet, spicy fragrance of the flowers is available for your olfactory pleasure from May to September.

This perennial's rhizomes form colonies in the limestone and caliche soils of the desert and open plains of eastern and central New Mexico, thriving from 4,000 to 7,500 feet elevation.

Like many members of the mint family, germander's leaves can be brewed into a fragrant herbal tea, and the plant also has been used to flavor alcoholic beverages. Germander tea is a diuretic and is recommended for gout and indigestion. The plant can be boiled down and the decoction rubbed on sore gums or on any wound to reduce inflammation. Germander was also a home remedy for expelling parasitic worms. It should be noted, however, that frequent use of germander may damage the liver.

Cut-leaf germander is a perfect border plant for a xeriscape or natural landscape planting. It blooms cheerily in very droughty conditions but will also respond favorably to extra moisture. It even blooms when growing in partial shade. It propagates by seed and root division as well as by rhizomes.

LILIACEAE (Lily Family)

Calochortus nuttallii

SEGO LILY, MARIPOSA LILY, BUTTERFLY TULIP

This handsome state flower of Utah is aptly named *Calochortus* (beautiful grass), simultaneously taking into account the stunning flower and the long, grasslike leaves. The species name honors the British botanist Thomas Nuttall. The common name "sego" is derived from the Southern Paiute word *sigho'o*, meaning "edible bulb." The resemblance of the petals to the wings of a butterfly (*mariposa* in Spanish) is acknowledged in the common names "mariposa lily" and "butterfly tulip."

Also shown here is another species, *Calochortus gunnisonii*, commonly called Gunnison's mariposa lily. The two *Calochortus* species differ slightly in their floral markings, with *Calochortus nuttallii* having a more

distinct cranberry "eyebrow." *Calochortus gunnisonii* is distinguished by its larger size and petal bases, which are cranberry-colored. Both plants bloom from May to July in overlapping ranges in northern, central, and western New Mexico in open grassland, sagebrush, and ponderosa pine communities from 4,500 to 11,000 feet elevation.

Like many members of the lily family, they have an underground bulb, which is one to two inches in diameter and tastes like a potato, or some say a peanut. This bulb was sought after for food by the early peoples of the Southwest. The Gosiute shared knowledge of sego lily's food value with the early Mormon settlers when they were starving from drought and locust invasions in the mid-1800s. Because of this blessing, the plant was named the state flower of Utah.

Calochortus gunnisonii, Gunnison's mariposa lily; whole plant. Note long, narrow grasslike leaves.

Calochortus gunnisonii, Gunnison's mariposa lily; three-chambered seedpod.

The Hopi and Navajo collected the bulbs in early spring before the plants bloomed. They peeled them and ate them raw, or cooked them by steaming, baking, or roasting in hot ash pits. The Navajo name for sego lily means "children's food," because it was one of their favorites. The Ute and Paiute usually waited until July to harvest the bulbs. They also relished the raw flower buds, and they ground the seeds into meal.

Calochortus is counted among the Life Medicines of the Navajo, who pulverized the leaves so that the juice could be applied to pimples. A decoction of the whole plant was given to women after childbirth to help expel the placenta.

Sheep and cattle graze on sego lily, and gophers, bears, and other mammals use the bulbs for food. Sego lily is an exceptionally attractive ornamental and grows best in well-drained sandy soil.

LOASACEAE (Stickleaf Family)

Nuttallia nuda (syn. *Mentzelia nuda*)

SAND LILY, BLAZING STAR

The genus name honors Thomas Nuttall, a nineteenth-century explorer and plant collector. The species name, *nuda,* means "naked."

This perennial is found in northern, far western, and far eastern New Mexico, growing in sandy soil in grasslands and along roadsides up to

Nuttallia nuda, sand lily; note young bud alongside candlelike seedpod.

about 6,000 feet elevation. The leaves and stems are covered by short, tenaciously gripping hairs that feel like Velcro when they grasp your clothes.

The flowers are almost four inches wide, with ten or more ostentatiously long white petals; they open late in the afternoon. The open flowers are replete with long, frilly white stamens, many of which are sterile, with no anther sac at the tip. Beneath each flower are large bracts that resemble many-lobed leaves. The flowers bloom from July to September and are followed by candlelike seedpods with what looks like a wick in the center (see photo); this is the remnant of the style.

The Dakota boiled the sap of sand lily and rubbed it on a feverish person to bring down the temperature.

For information about another *Mentzelia* species, see p. 110.

NITRARIACEAE (Harmal Family)

(formerly in Zygophyllaceae)

Peganum harmala

AFRICAN RUE, SYRIAN RUE, HARMAL

The story of the human desire to dress in red is one of economic intrigue and competition, with parallel stories of ecological calamity and human suffering. The monopoly of the cochineal trade by the Spaniards following their discovery of the cochineal scale insect on *Opuntia* and *Cylindropuntia* cacti in the New World is possibly the most dramatic tale (see p. 301). More recently, in the twentieth century, New Mexico became ground zero for a noxious weed introduced from the Old World into the southwestern states in pursuit of an alternative source of "Turkey red" dye.

During the 1700s and 1800s, the textile industry in Europe was looking for a bright, colorfast dye for cotton to supply cloth to the growing colonial markets. The red dye, alizarin, made from the root of madder (*Rubia*), was arduously produced in Turkey (hence the name "Turkey red"). On a smaller scale, in the Middle East, India, and North Africa, a red dye was extracted from the seeds of *Peganum harmala*, the plant shown here.

In the 1920s, a farmer from Deming, New Mexico, imported seeds of *P. harmala* with the intent of extracting the red dye from them for coloring wool. The result was the inadvertent and rapid spread of this invasive (and toxic) plant throughout the Southwest, where most states, including New Mexico, now classify it as a noxious weed.

African rue contains toxic alkaloids concentrated in the seeds and fruits. In Asia and the Middle East, it has a long history of use as a medi-

Peganum harmala, African rue; developing fruit.

cine for eye problems, parasitic worms, insomnia, viral diseases, and tumors. It is also used as an aphrodisiac, to increase milk production, and to induce abortion. One of its chemical constituents is harmaline, a nervous system stimulator that has been used to treat depression. It is also exploited for its psychotropic effects. The plant also contains vasicine, which is effective in killing the protozoan that causes the tropical disease leishmaniasis.

The plants are very unpalatable, but animals may eat them if they have no other food or have a mineral deficiency. Cattle and sheep are adversely affected by even a small amount of the plant.

NYCTAGINACEAE (Four O'clock Family)

Abronia fragrans

FRAGRANT SAND VERBENA, SWEET SAND VERBENA, SNOWBALL SAND VERBENA

Sand verbena is not a verbena at all, but the arrangement of the flowers in a globose cluster could trick someone into thinking it was. The scientific

Abronia angustifolia, purple sand verbena; flower with developing seedpod at White Sands National Monument.

name, *Abronia fragrans*, meaning "delicate" and "fragrant," corroborates the adjective components of the common names.

This perennial grows throughout New Mexico in sandy soil from 4,000 to 6,000 feet elevation. The plant is somewhat erect, reaching two or three feet tall, but also tends to recline on the ground. Typical of the four o'clock family, the leaves are opposite each other on the stems and are unequal in size. The entire plant is covered with sticky hairs. The flowers bless their environment with sweet smells from April through August, opening in late afternoon (around four o'clock!) and remaining open until the next morning. After pollination, the flowers evolve into a spiky, extraterrestrial-looking fruiting structure before finally maturing into clusters of small fruits (achenes) with five wings (see photos).

The Navajo and Zuñi had colorfully descriptive names for fragrant sand verbena, including "stinkbug food," "spider medicine," and "stomachache medicine." They ground the starchy roots and mixed them with cornmeal. They believed these roots would enhance a person's appetite and help fatten him or her up. They wore the flowers as a necklace during ceremonies.

Listed among the Life Medicines of the Navajo, sand verbena has many useful qualities: it is a sedative, a sudorific (sweat inducer), a

Abronia angustifolia, purple sand verbena; "extraterrestrial" immature seedpod.

Abronia angustifolia, purple sand verbena; mature gossamer-winged seedpod.

diuretic, and an antispasmodic. The Kayenta Navajo used it to treat insect bites, to calm stomach cramps, and to empty the digestive tract. The Ramah Navajo made an infusion to wash sores and perspiring feet, while the inhabitants of Chaco Canyon used it to treat boils.

In New Mexico, some Hispanic women drank *A. fragrans* tea every third day to stimulate the flow of breast milk and used the tea to bathe their breasts.

Abronia species yield various shades of brown dyes for wool from all parts of the plant, with the final color dependent on the type of mordant used to fix the color.

Fragrant sand verbena is exceptionally desirable as an ornamental in xeriscapes or rock gardens, not only perfuming the atmosphere but also attracting night-flying moths.

Also shown here is a purple-flowered cousin, *A. angustifolia,* found throughout the southern half of New Mexico and a beautiful feature among the flora of White Sands National Monument. The plant is typically an annual, but it may be a perennial on the gypsum flats of White Sands.

ONAGRACEAE (Evening Primrose Family)

Oenothera pallida subsp. *gypsophila*

PALE EVENING PRIMROSE, WHITE EVENING PRIMROSE

The species name, *pallida* (pale), aptly describes the white flowers. The subspecies name, *gypsophila* (gypsum lover), tells us that this plant is able to grow in alkaline gypsum soil, such as that of White Sands National Monument, where this photo was taken.

Pale evening primrose, a perennial that spreads by rhizomes, reaches up to two feet tall and has reddish stems. The leaves are generally long and narrow. It is found throughout New Mexico and grows in a variety of soil types, including gypsum.

Evening primrose leaves can be eaten as a cooked vegetable, the roots can be boiled and eaten, and the seeds are also edible.

In preparation for certain ceremonies, the Navajo purified themselves with pale evening primrose, which is a purgative. They boiled the roots and leaves to make a liniment for snakebites and made a poultice of the plant for spider bites. The Navajo drank a tea to treat urinary tract problems, sore throat, and upset stomach. This infusion was also given to livestock with colic. Venereal disease sores were dusted with a powder made from the dried plant.

All *Oenothera* species are superlative choices as landscaping features, especially in rock gardens, where they grow well in poor or dry soil. Evening primroses are pollinated by moths and visited by a number of other insects. Deer and other animals eat the leaves, while birds and small rodents consume the seeds.

Oenothera albicaulis, prairie evening primrose; note X-shaped stigma typical of *Oenothera* genus.

Oenothera coronopifolia, crownleaf evening primrose; note thick hairs at base of anthers.

Two other white evening primroses shown here are *Oenothera albicaulis,* whose flowers remain white as they age and which grows throughout New Mexico, and *O. coronopifolia,* whose flowers turn pink with age and which grows in the northern and western halves of the state.

For information about other members of the evening primrose family, see pp. 16, 111, and 113.

PAPAVERACEAE (Poppy Family)

Argemone squarrosa

HEDGEHOG PRICKLY POPPY

Argemone (from the Greek *argemos,* "a white spot, or cataract, on the eye") was the name used by Dioscorides for a poppy that was used to treat cataracts. However, the American *Argemone* has a different alkaloid profile, and its use for the eyes has unfortunately resulted in glaucoma. The species name refers to the spreading spines on the bracts underneath the seedpods.

Argemone squarrosa, hedgehog prickly poppy; prairie view of whole plant.

Argemone squarrosa, hedgehog prickly poppy; young, very prickly seedpods.

Hedgehog prickly poppy is a three-foot-tall perennial growing in eastern New Mexico in dry prairies and foothills from 3,000 to 6,000 feet elevation. The handsome flowers are well protected by incredibly well-armed body parts, including leaves, stems, buds, and fruits (see photo).

Powerful alkaloid toxins, including berberine, in the yellow latex of *Argemone* strike it from any list of potential food sources but make it a great medicine. Native Americans used the pulverized seeds and the sap to make liniments for burns, sores, warts, and cuts. They made a paste from the seeds to rub in the hair to kill head lice. They warmed the roots to poultice the jaws of toothache sufferers. Medical practitioners among the early settlers used it to treat skin diseases and relieve colic. Herbal healers in Mexico still use it for a host of medical problems, including asthma, liver disease, and bad breath. The Comanche and Shoshone collected the colored sap to use as a salve for sore eyes, but they likely inflicted worse damage than the original ailment. The Tepehuan people of northwest Mexico crushed the plant and sprinkled it in their fishing waters to stun the fish for easy capture. The Kiowa put ashes from *Argemone* under the skin to create tattoos, and the Lakota colored their arrows yellow with a dye made from the plant.

The small seeds provide food for quail and doves. Since cattle avoid eating this plant, overgrazing will result in increasingly large colonies of hedgehog prickly poppy and degradation of pastureland.

POLYGONACEAE (Buckwheat Family)

Eriogonum jamesii

JAMES BUCKWHEAT, ANTELOPE SAGE

One of the traits of the buckwheat family is "hairy knees," which is the literal meaning of the genus name, *Eriogonum*. The species name recognizes Edwin James, a nineteenth-century plant collector with the Long Expedition to Colorado in 1819–20.

There are more than forty species of *Eriogonum* in New Mexico, almost all with finely narrow stems that fork dichotomously (in pairs) and papery flowers that are grouped in clusters. This perennial grows on dry rocky slopes in grasslands and pinyon-juniper communities up to about 7,000 feet elevation. It blooms from late summer to fall with clusters of cream-colored flowers on foot-tall stems. The leaves and flowers are hairy.

Carbonized buckwheat seeds have been unearthed at several archaeological sites, indicating their prehistoric use as food. In fact, the plants still grow in old gardens and agricultural areas around ruins. The practice of eating buckwheat has continued among the descendants of the original inhabitants, with the Navajo eating the roots and grinding the seeds into a meal for porridge.

Eriogonum jamesii is a Life Medicine for the Navajo, whose women relied on a decoction of antelope sage to ease labor pains. And if they did not wish to go through that again, they prevented conception by drinking a solution made from the boiled roots. The Acoma and Laguna chewed the roots to strengthen the heart and drank a root infusion to brighten their mood when feeling depressed. The Kayenta Navajo smoked antelope sage when they had disturbing dreams, and the Zuñi used powdered buckwheat flowers in rain ceremonies.

Eriogonum alatum, winged wild buckwheat; note three-winged seeds.

Eriogonum jamesii provides food for both larvae and adults of western square-dotted blue (*Euphilotes battoides*), blue copper (*Lycaena heteronym*), and Mormon metalmark (*Apodemia mormo*) butterflies. It is a nectar source for the rare Spalding dotted-blue butterfly (*Euphilotes spaldingi*).

Also pictured here is *Eriogonum alatum* (note winged seed in photo), which is found in the same habitat as *E. jamesii* throughout the state.

ROSACEAE (Rose Family)

Cercocarpus montanus

MOUNTAIN MAHOGANY

The fruits (achenes) of mountain mahogany have a feathery corkscrew plume (see photo) that gives the genus its name, *Cercocarpus*, meaning

Cercocarpus montanus, mountain mahogany; fuzzy-tailed fruit at Bandelier National Monument.

"tail fruit." The species name, *montanus*, or "mountain," gives an indication of where it grows, from 6,000 to 9,000 feet elevation.

This shrub is usually four to six feet tall and blooms in April and May. It grows throughout New Mexico on rocky slopes in chaparral and woodland communities, as shown here at Bandelier National Monument.

The Navajo drank an extract of mountain mahogany when they ate too much (an early Alka-Seltzer!), and the pulverized leaves were sprinkled in water and drunk for a laxative. An extract was also given to women to help them recover from childbirth. The Kawaiisu of California boiled the roots and drank the decoction as cough medicine.

The tannins in the bark were useful in tanning leather, and the roots provided a red dye for wool, baskets, and buckskin.

The wood produced by this small tree is brittle but very hard. The Acoma and Laguna used the tough, thick branches to fashion fire stick spindles. The people of several pueblos used the wood to make arrow points, digging tools, spears, rabbit sticks (see photo), and rough brooms. Navajo rug makers still value the wood to make weaving tools.

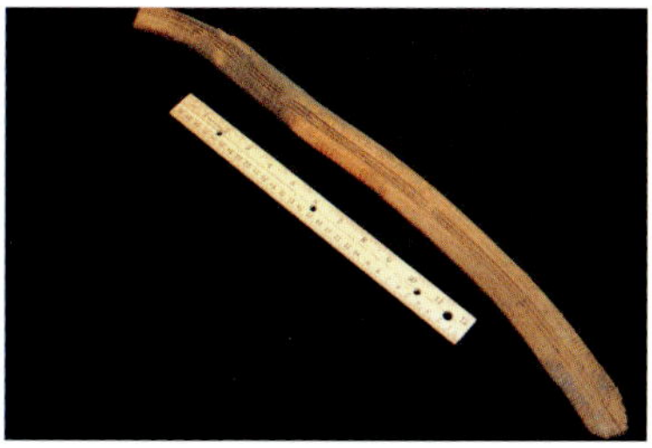

Rabbit stick, used like a non-returning boomerang for hunting rabbits. Artifact found in cave on Lower Pecos River.

Several wildlife species, including deer and elk, use mountain mahogany for some or much of their diet at different times of the year. The plant rarely achieves its full growth potential because of constant browsing. Small mammals and a wide variety of birds use it for both feed and cover, while coyotes and collared peccaries use it for cover.

Mountain mahogany has the potential to be toxic to livestock, especially ruminants, if they consume damaged or wilted leaves. The plant contains cyanogenic glycosides that break down to release cyanide when the leaves are damaged, but not always enough to be dangerous. Cyanide acts very quickly, and death can occur within fifteen minutes.

Cercocarpus montanus has excellent potential in xeriscapes or natural landscapes in dry, well-drained sites. It is also useful in revegetation or reclamation projects, where the plants stabilize the soil and fix nitrogen to encourage plant growth. Both seed and plants are commercially available.

Mountain mahogany hosts caterpillars of the Behr's hairstreak butterfly (*Satyrium behrii*) and the western sheepmoth (*Hemileuca eglanterina*).

ROSACEAE (Rose Family)

Fallugia paradoxa

APACHE-PLUME

The genus name honors Virgilio Fallugi, an Italian monk and botanist of the 1600s who lived near Florence. The species name, *paradoxa,* means "contrary to expectations," referring to the plant's un-rose-like appearance. The pink-purple plumes are reminiscent of Apache war bonnets, hence the common name.

Apache-plume is a six-foot-tall evergreen shrub with wedge-shaped, deeply divided leaves. It makes its home throughout New Mexico in several communities, including grasslands and pinyon-juniper, from 3,000 to 8,000 feet elevation. As a phreatophyte, it has deep roots that can access subsurface water, giving it a special strategy for surviving desert conditions. The flowers appear from April through October, and as the

Fallugia paradoxa, Apache-plume; whole plant.

fruits mature, the styles remain attached, creating a feathery show on the bushes through the summer and fall and into the winter.

When Native Americans suffered pain or fever, the flowers and bark of Apache-plume provided relief. Those with joint pain would place a poultice of the leaves with tobacco on the aching body part. Poultices made of the plant's beautiful white flowers were placed on swollen legs (of humans or horses) to reduce distension. The Keresan people of the Cochiti Pueblo near Santa Fe ate the flower petals to prevent stomach gas, while others ate the tender stems to control indigestion.

The Ramah Navajo drank an infusion of the leaves as a ceremonial emetic for internal cleansing, and they used the same solution in a purifying bath. The straight stems were used as arrow shafts, while the strong, flexible roots served as cording to tie fences. One of the most common uses of the stems today is for yard brooms, made by tying together many small branches.

Apache-plume is very drought tolerant and, with its showy flowers and fruits, makes an exceptional addition to a xeriscape. It is useful in rehabilitation of disturbed sites for erosion control and in stabilizing soil in arid environments because it spreads easily from rhizomes. Both seed and plants are available commercially.

Although it is considered mediocre-quality browse for livestock and wildlife, Apache-plume provides good cover for small mammals and birds.

ROSACEAE (Rose Family)

Purshia stansburyana

CLIFFROSE, QUININE BUSH, STANSBURY CLIFFROSE

The scientific name, *Purshia stansburyana*, honors two men with connections to plant collections from the West. The genus name recog-

Purshia stansburyana, cliffrose; scenic view of whole plant.

Purshia stansburyana, cliffrose; plumed tails on seed.

nizes Frederick Pursh, an eighteenth- and nineteenth-century German American botanist who published a book about plants collected during the Lewis and Clark Expedition. The species name honors Howard Stansbury, a nineteenth-century American who made extensive studies of Great Basin flora and fauna.

Cliffrose, a six-foot-tall evergreen bush, does indeed inhabit cliffs, most often in limestone-derived soil in many plant communities from 5,000 to 8,000 feet elevation in western New Mexico. Each midget leaf has stubby little "fingers" and is covered with glands that release a pungent, quinine-like odor when crushed, leading early pioneers to call the plant "quinine bush." The April-to-June bloom period is followed by fruit set, heralded by five long plumose tails (see photo) on the hairy achenes, or "shrink-wrapped" seeds.

The aromatic triterpenoids in cliffrose leaves are used by the Navajo for coughs and as a wash for wounds and other skin problems. These triterpenoids have been investigated for their inhibitory effects on HIV and the Epstein-Barr virus.

The Navajo used the absorbent, shreddy bark to pack around their babies' bottoms as diapers; after being dried in the sun, the bark was reused. The cushiony bark had many other uses as well: as comfortable padding around infants in their cradleboards, as filling for pillows, and as fiber in making clothing, mats, and rope (see photo).

Example of plant fiber rope; artifact found in cave on Lower Pecos River.

Navajo women made prayer sticks from the hard wood, while Acoma and Laguna men made arrow shafts from its strong, straight stems.

Traditional Navajos still make an offering of a perfect grain of white corn placed in the bark of cliffrose to help them correct wrong thoughts that are expressed as bad dreams. Navajo weavers extract brown and yellow dyes from this plant.

Cliffrose is important as both browse and cover for many wildlife species, including deer, bighorn sheep, elk, pronghorn, many game birds and songbirds, and small mammals. Livestock also browse it heavily in winter.

Cliffrose is a colonizer on disturbed sites, such as after a fire or mining activities, and is recommended for habitat restoration. It fixes atmospheric nitrogen, which helps to improve soil fertility in reclamation projects. It is also marketed as an ornamental for its drought-tolerant attributes and its showy, fragrant flowers. A variety of insect life is attracted to the flowers, and cliffrose is a larval host for the desert elfin butterfly (*Callophrys fotis*).

SANTALACEAE (Sandalwood Family)

Comandra umbellata

BASTARD TOADFLAX

Comandra has enjoyed its genus name since Thomas Nuttall created it in 1818. The name means "male hairs," referring to the hairy stamens. *Umbellata* describes the "umbrella-like" arrangement of the flower clusters. The common name, while very colorful, is not so informative.

Comandra umbellata is a hemiparasite like *Castilleja* (see p. 18), surreptitiously mooching nutrients from its neighbors by attaching its roots to theirs and shuttling food and water to home base. Meanwhile, the leaves do have chlorophyll, with which the plant can produce its own carbo-

hydrates. The plant is a perennial and grows throughout New Mexico in dry, rocky, or sandy soil in desert (as in this photo from White Sands National Monument) to pinyon and ponderosa pine zones up to 8,000 feet elevation. Flowering season is April through June.

Though not a major food source, bastard toadflax seeds and fruits are nutritious and have been eaten by Native Americans. Children also enjoy sucking the sugary nectar from the flowers.

The plant's juice was applied to sores and wounds, and a tea made from the leaves was drunk to clear congested lungs. The tea was dropped into irritated eyes and swished in the mouth to heal canker sores. The Kayenta Navajo also valued bastard toadflax as a narcotic.

Comandra umbellata is a minor accumulator of selenium and can cause toxicosis, especially in horses. A light blue dye is extracted from the roots.

Butterflies are attracted to the flowers for their nectar, and the plant hosts caterpillars of the common buckeye butterfly (*Junonia coenia*).

SOLANACEAE (Nightshade Family)

Datura wrightii (syn. *Datura ferox*)

SACRED DATURA, SPINY APPLE, JIMSONWEED

The genus name of this powerfully dangerous *Datura* comes from *Dhatura*, the Sanskrit name for a species found in India and used in rituals similar to those of Native Americans. Charles Wright, honored with the

Datura wrightii, jimsonweed; cutaway of mature spiny pod with seeds.

Datura wrightii, jimsonweed; corolla in process of unfurling.

species name, was a nineteenth-century American botanical explorer and collector in the American West.

Spiny apple has a striking appearance, with six- to eight-inch-long, trumpet-shaped flowers that unfurl at the end of each day to await visits from nocturnal hawk and sphinx moths, whose very long proboscises are perfectly suited to plumb the depths of the narcotic-laced nectar. These compounds apparently intoxicate the moths, which become discombobulated after imbibing.

This three-foot-tall, bushy annual grows in disturbed soil from 4,000 to 6,000 feet elevation and blooms from May through November. The seedpods are covered with sharp, heavy spines and are known as "spiny apples." A related species, *D. ferox,* has seedpods that are "ferociously fortified" (*ferox*), giving fair warning of the dangers held within.

The synonymous name of *Datura ferox, D. quercifolia,* indicates that the plant has leaves that resemble those of Gambel oak (*quercifolia* means "oak leaves"). *Datura quercifolia* flowers are tinged with violet and are not as

Datura quercifolia, oak-leaf thornapple; note elongated, angular calyx encircling floral throat.

Datura quercifolia, oak-leaf thornapple; spinescent young seedpod.

fully unfurled as those of *D. wrightii,* which is found in all regions of New Mexico, whereas *D. quercifolia* is found only in the central counties.

Datura wrightii emits a musty, narcotic-like smell. All parts of the plant are deadly toxic, with the belladonna alkaloids, including atropine, hyoscine, scopolamine, and hyoscyamine, as the agents of death. The seeds concentrate these toxins and are more deadly than the other parts. The combination of anticholinergic compounds renders a person unable to differentiate between fantasy and reality. The psychoactive properties of *Datura* have associated plants of this genus with magic and witchcraft more than any other. In the Navajo heritage, it is seen as an evil plant because it was used long ago by a powerful evil individual known traditionally as "the Gambler" to drug and enslave people of the Chaco Canyon Culture, leading to the breakup and dispersal of the ancestors of the modern-day tribes.

One should never experiment with such a dangerous plant, because the potency of the toxins varies greatly from plant to plant and depends on location and growing conditions, as well as season. Even experienced people have made fatal mistakes when trying to estimate proper dosages. Livestock are as much at risk from the effects as humans and will die after consuming only a few ounces of the plant.

Despite the dangers of direct consumption of *Datura,* Native Americans used it in vastly smaller doses for several medical applications. They crushed the seeds to make an ointment to exterminate body lice, breathed in the smoke from burning leaves during asthma attacks, and used the powdered roots as an anesthetic during painful procedures such as sewing up wounds. The Acoma and Laguna crushed the leaves or roots and placed them on boils. After the Navajo castrated young sheep, they washed the cuts with an infusion of *Datura* leaves. This liquid was also used as an eyewash for blindness in horses and a cleanser for sores.

Adult sphinx and hawk moths, including pink-spotted hawkmoth (*Agrius cingulata*), five-spotted hawkmoth (*Manduca quinquemaculata*), and Carolina sphinx (*Manduca sexta*), nectar on *Datura* flowers. Their caterpillars feed on the foliage.

SOLANACEAE (Nightshade Family)

Lycium pallidum

PALE WOLFBERRY, DESERT THORN, TOMATILLO

A thorny shrub from the Lycia region of Turkey, named *Lykion* by Dioscorides and Pliny about two thousand years ago, is the source of the genus name, *Lycium*. The common name "wolfberry" probably comes from the presumed derivation from *lykos*, or "wolf." The species name, *pallidum*, means "pale," referring to the light-colored flowers.

This six-foot-tall, spiny shrub has entangled branches covered with leathery leaves that fall off when conditions are too hot and dry. When the rains arrive, however, the shrub will releaf.

Pale wolfberry grows in limestone soil in several plant communities from 3,500 to 7,000 feet elevation in the Chihuahuan and Sonoran Deserts, as well as throughout central, western, and northern New Mexico. The flowers appear in May and June, followed by bright orange or red berries, which disappear quickly into the stomachs of hungry humans and wildlife, especially birds and rodents.

The Ancestral Puebloans relied on wolfberries as a food source, as do modern-day tribes in the Southwest. The Hopi eat the small orange-red berries fresh from the shrub or cook them into a jam to serve with piki bread. The Navajo consider wolfberries as a "food of the gods" and eat them in porridge as part of religious ceremonies rather than as regular food.

The Navajo count *Lycium* among their Life Medicines and used pale wolfberry for various remedies. They ground the root to place inside cavities to combat toothache. They also treated chicken pox sores with preparations of the plant. The Rio Grande Puebloans applied poultices of the leaves to cuts.

The Ramah Navajo use wolfberry as a ceremonial emetic to induce vomiting, and they blacken themselves with ashes from the burned plant. The Zuñi historically gave the powdered leaves, twigs, and flowers to warriors prior to battle, and they still use the plant at the annual Niman Kachina Ceremony.

Archaeologists use pale wolfberry as an indicator plant to detect middens of the Ancestral Puebloans, because it tends to grow where they once lived and ate the berries, spreading the seeds around their home sites and possibly even cultivating them. The hard-packed floors of the now buried prehistoric rooms hold rainwater, and this extra moisture encourages more luxuriant growth of the plant over ruins. A straight row of bushes generally indicates a buried wall.

Wolfberry provides food and cover for birds and small mammals. It is an attractive member of a xeriscape and also makes a good barrier hedge, with its complement of thorns. The tough and extensive root system can extend up to thirty feet from the plant. Plants can be propagated by cuttings and layering (covering a branch with soil to stimulate root growth at the nodes).

achene A small, dry, usually hard one-seeded fruit that does not split open at maturity. Example: a sunflower seed.

alkaloid An organic compound produced by plants that contains nitrogen. Many alkaloids are toxic but may have medicinal applications.

allelopathic Producing chemicals that inhibit the growth of neighboring plants.

alpine Pertaining to mountains, particularly the uppermost parts beyond the tree line.

alternate leaf arrangement One leaf per node, alternating positions on the stem so as not to block the sun from the leaves directly beneath them.

annual Completing the life cycle in one year or growth season.

anodyne Anything that relieves or soothes pain.

anther A pollen sac at the end of a stamen.

anticholinergic Any substance (such as atropine) that inhibits the transmission of nerve impulses, thereby decreasing spasms in smooth muscles.

antispasmodic Preventing or relieving spasms.

astringent Causing the constriction and shrinkage of body tissue and blood vessels, checking blood flow.

atropine A poisonous alkaloid with medical applications, such as dilating pupils, preventing spasms, and drying up secretions. Produced by some members of the nightshade family (Solanaceae).

bajada A fan-shaped deposit of debris at the base of a mountain slope.

basal At the base or bottom of a plant.

biennial A plant whose life cycle takes two years to complete.

bilateral symmetry Having two sides that are mirror images of each other.

bipinnate Doubly pinnate, with both the primary and secondary divisions of a compound leaf arranged in a featherlike, or pinnate, fashion.

berry A pulpy fruit with no stone. Example: a tomato.

bract A leaf associated with a flower or group of flowers. Some are colored, mimicking petals. Example: the red "petals" of Indian paintbrush.

bulb An underground bud with overlapping, thickened leaves. Example: an onion.

calcareous Containing much calcium, usually in the form of calcium carbonate; limy.

caliche A crusty, hard layer of calcium carbonate on certain soils in arid regions.

calyx A collective term for the sepals. Found under the petals, they sometimes resemble leaves and sometimes resemble petals. The calyx encloses and protects other flower structures in the bud.

campanulate Shaped like a bell.

cathartic An extreme laxative.

catkin A scaly, petalless inflorescence with flowers all of one sex. Example: a group of flowers on a willow.

climax community A mature assemblage of flora and fauna that has reached a steady state over time.

community An assemblage of flora and fauna of various species inhabiting a particular region.

compound leaf A leaf in which the blade is divided into individual leaflets.

connate Fused; united at the base in pairs around an axis.

coprolites Fossilized feces; studied by archaeologists to determine diet.

corm An underground stem, short and rounded, that is used for food storage and asexual reproduction.

corolla A collective term for all the petals of a flower.

decoction An extract obtained by boiling down plant parts.

dehiscent Splitting open at maturity.

demulcent Having a softening or soothing effect.

dioecious Having male and female reproductive structures separated on two different plants.

diploid Having two sets of chromosomes, one set from each parent.

disk flower A small, tubular flower in the center of a flower in the Asteraceae (sunflower) family. Example: the individual brownish flowers in the center of a sunflower.

diuretic Increasing urination, thus removing excess fluid.

edema The state of being swollen with excess fluid.

elaiosomes Oil- and protein-rich bodies on certain seeds used by ants for food.

emetic Causing vomiting.

emollient Soothing, softening, making supple.

expectorant A substance that helps a person cough up phlegm and spit it out.

fascicle A small bundle.

febrifuge Fever reducing.

fumigant A substance used to produce smoke or fumes.

fusiform Shaped like a spindle, narrow at the ends and wider in the middle.

genus The first name in the two-part scientific name. Very closely related species are in the same genus.

halophyte A plant that grows in salty soils.

haustoria Suckerlike attachment organs of parasitic plants.

hemiparasite A plant that is a partial parasite, making its own food but taking substances from a neighboring plant through specialized root structures called haustoria.

indehiscent Not splitting open at maturity.

inflorescence An arrangement of flowers on a stem or along an axis; a group of flowers on a plant.

infusion A liquid, or "tea," obtained by soaking plant parts in water so as to extract certain compounds.

melittophilous Pollinated by bees.

monoecious Having male reproductive parts and female reproductive parts in separate structures but on the same plant. Example: a pinyon pine tree, which has both male and female cones on the same tree.

myrmecochory Seed dispersal by ants.

nixtamalization The process of soaking corn in an alkaline solution to release the hull and improve the protein and vitamin value while also reducing the toxins from fungi.

node The point on a stem where leaves and buds are attached.

opposite leaves A pair of leaves that grow directly across from each other at a single node.

ornithophilous Pollinated by birds.

ovary The part of the pistil (the female part of a flower) where ovules are formed. Upon maturity, the ovary becomes a fruit and the ovules become seeds.

palmate Appearing in form like an open hand with the fingers extended. The term can refer to the arrangement of veins in a leaf that fan out from the base, or to the arrangement of leaflets in a compound leaf in which all attach at one point to the petiole.

pappus A crown of bristles or scales on the achene (seed) of a member of the sunflower family (Asteraceae). Example: the fluffy down of a mature dandelion head.

peduncle The stalk supporting a single flower or a flower cluster.

perennial A plant whose life cycle persists for more than two years.

petiole The stalk connecting a leaf to the stem.

phreatophyte A plant whose deep roots are constantly in contact with groundwater.

phyllary One of a group of leafy bracts beneath the flower head of a composite plant.

pinnate Shaped like or resembling a feather; may refer to the arrangement of veins in a leaf in which small veins branch off one central main vein, or to the arrangement of leaflets in a compound leaf in which the small leaflets are attached along either side of a common rachis.

pinnatifid A pinnately lobed leaf, with lobes divided more than halfway to the central vein.

piscicide A fish poison.

pistil The female portion of a flower, consisting of a stigma, style, and ovary.

ploidy level The number of chromosome sets in an organism's individual cells. The normal number is two sets (diploid), but some plants may have more (triploid, tetraploid, etc.).

plumose Featherlike.

poultice A hot, moist mass applied to an injured or diseased part of the body.

pubescent Downy; covered with soft hair.

purgative An extreme laxative.

raceme An elongated inflorescence in which each flower is attached by a short stalk to a common stem.

rachis The axis of a raceme or a compound leaf.

ray flower Each individual small flower that mimics a petal in a flower of the sunflower family (Asteraceae). Example: the yellow "petals" of a sunflower are actually ray flowers.

rhizome An underground stem that gives rise to roots and shoots.

riparian Growing along the banks of a river.

scarify To make small scratches or incisions in a seed coat to aid water absorption for germination.

sepal A leaflike structure that encloses and protects the floral parts in the bud. After the bud opens, the sepal remains under the flower. Usually green, it is sometimes colored, resembling a petal.

seral community An intermediate stage in an ecosystem's development toward a climax community.

simple leaf A leaf in which the blade is not divided into leaflets.

stamen The male structure of a flower, made up of the pollen-bearing anther and a supporting filament.

staminode A sterile stamen.

stanch To stop or check the flow of blood.

stellate Star-shaped.

stigma The receptive end of the female structure, or pistil, of a flower, where pollen lands and germinates.

stipule A small, leaflike appendage at the base of a leaf stalk, or petiole.

style The narrow portion of the female structure, or pistil, connecting the stigma and ovary.

sudorific Increasing perspiration.

tannin Tannic acid used in tanning hides and in medicine. Obtained from the roots, bark, sap, and fruits of certain plant species.

tea An infusion obtained by soaking plant parts in water.

tetraploid Having four sets of chromosomes rather than the normal two (diploid).

tonic A medicinal substance that generally invigorates the health.

trichome A hairlike outgrowth of a plant's epidermis.

trifoliate Having three leaves.

trifoliolate Having three leaflets.

tuber A thick, solid, rounded underground stem used for food storage and for asexual reproduction. Example: a hog potato tuber.

umbel An inflorescence in which the flower stalks rise from a common point, resembling an umbrella.

vegetation zone An area comprising plant species typically associated together at a particular elevation and named for a dominant woody species. Example: the pinyon-juniper zone is from 5,000 to 8,000 feet elevation.

vermifuge A medicine used to rid the body of parasitic worms.

vespertine Blooming in the evening.

whorled leaves Three or more leaves attached at a single node.

xeriscape A landscaping method that eliminates the need for supplemental watering; desert landscaping.

Agricultural Network Information Center. A Knowledge Discovery System for Agriculture. Rhodes University Libraries. www.agnic.org/search/IND44326506 (accessed November 10, 2011).

Agricultural Research Service. U.S. Department of Agriculture. www.ars.usda.gov/Main/docs.htm?docid=9749 (site updated October 27, 2005; accessed November 10, 2011).

AgroAtlas. Interactive Agricultural Ecological Atlas of Russia and Neighboring Countries. Economic Plants and Their Diseases, Pests and Weeds. www.agroatlas.ru/en/content/weeds/ (accessed November 10, 2011).

Alexander, Patrick. Polyploid.net. http://polyploid.net/plants.html (accessed October 19, 2011).

Allred, Kelly W. 2010. "An Annotated Checklist of Poisonous or Injurious Range Plants of New Mexico." New Mexico State University. http://cahe.nmsu.edu/pubs/research/livestock_range/RR715.pdf (accessed December 14, 2011).

Arches National Park. National Park Service, U.S. Department of the Interior. www.nps.gov/arch/naturescience/search_by_name.htm (site updated May 27, 2010; accessed November 3, 2011).

Armstrong, Wayne. *An Online Textbook of Natural History*. San Marcos, CA: Life Science Department, Palomar College. http://waynesword.palomar.edu (site updated July 1, 2011; accessed October 19, 2011).

Arnberger, Leslie P. 1982. *Flowers of the Southwest Mountains*. Tucson, AZ: Southwest Parks and Monuments Association.

Austin, Daniel F. 2004. *Florida Ethnobotany*. Boca Raton, FL: CRC Press.

Australian Naturopathic Network. www.ann.com.au/herbs/Monographs/equisetu.htm (site updated June 2008; accessed November 15, 2011).

Beardsley, Paul M., and Richard G. Olmstead. 2002. "Redefining Phrymaceae: The Placement of *Mimulus*, tribe Mimuleae, and *Phryma*." *American Journal of Botany* 89: 1093–1102. www.amjbot.org/content/89/7/1093.abstract (accessed December 26, 2011).

Beasley, V. 1999. "Toxicants with Mixed Effects on the Central Nervous System (Part II)." In *Veterinary Toxicology*. Ithaca, NY: International Veterinary Information Service. www.ivis.org/advances/beasley/Cpt2d2/IVIS.pdf (accessed October 19, 2011).

Beauduy, Reed. 2004. "*Agave parryi* Engelm. var. *parryi*." Native Plants of Arizona. http://jan.ucc.nau.edu/~plants-c/bio414/species pages/Agave parryi.htm (accessed November 3, 2011).

Belcher, Earl, et al. 2003. *Handbook on Seeds of Browse-Shrubs and Forbs*. The

Browse-Shrub and Forb Committee of the Association of Official Seed Analysts. Atlanta: USDA Forest Service, Southern Region.
Bitter Rubberweed (*Hymenoxys odorata*). Poisonous Plant Research, USDA Agricultural Research Service. http://ucjeps.berkeley.edu/interchange/I_treat_indexes.html (site updated February 7, 2006; accessed October 27, 2011).
BLM (Bureau of Land Management). www.blm.gov/wo/st/en.html (site updated October 31, 2011; accessed November 3, 2011).
Botanical Dermatology Database. www.botanical-dermatology-database.info/index.html (site updated October 18, 2011; accessed November 5, 2011).
Bowers, Janice Emily. 1989. *100 Desert Wildflowers of the Southwest*. Tucson, AZ: Western National Parks Association.
Bowers, Nora Mays, Rick Bowers, and Stan Tekiela. 2008. *Wildflowers of Arizona Field Guide*. Cambridge, MN: Adventure Publications.
Bowns, James E. 1999. "Ecology and Management of Pinyon-Juniper Communities within the Interior West: Overview of the 'Resource Values Session' of the Symposium." *USDA Forest Service Proceedings* RMRS-P-9. www.fs.fed.us/rm/pubs/rmrs_p009/rmrs_p009_157_163.pdf (accessed December 26, 2011).
Bridges, M., A. M. E. Jones, A. M. Bones, C. Hodgson, R. Cole, E. Bartlet, R. Wallsgrove, V. K. Karapapa, et al. 2002. "Spatial Organization of the Glucosinolate-Myrosinase System in *Brassica* Specialist Aphids Is Similar to That of the Host Plant." *Proceedings of the Royal Society B: Biological Sciences*. http://rspb.royalsocietypublishing.org/content/269/1487/187.short (accessed November 22, 2011).
Brown, Walter C., Angel C. Alcala, Arvin C. Diesmos, and Ely Alcala. "Species of the *Guentheri* Group of *Platymantis* (Amphibia: Ranidae) from the Philippines, with Descriptions of Four New Species." *Proceedings of the California Academy of Sciences* 50 (1): 1–20. www.archive.org/stream/proceedingsofcal0450cali#page/n5/mode/1up (accessed December 26, 2011).
Burdock. Complementary Medicine. Medical Reference. University of Maryland Medical Center. www.umm.edu/altmed/articles/burdock-000227.htm (accessed November 4, 2011).
Burow, M., A. Bergner, J. Gershenzon, and U. Wittstock. 2007. "Glucosinolate Hydrolysis in *Lepidium sativum*—Identification of the Thiocyanate-Forming Protein." *Plant Molecular Biology* 63 (1): 49–61. doi:10.1007/s11103-006-9071-5. PMID 17139450 (accessed November 22, 2011).
Cabeza Prieta Natural History Association. http://cabezaprieta.org/index.php (accessed November 3, 2011).

Calflora. www.calflora.org/ (accessed November 3, 2011).
California Native Plant Wiki. Theodore Payne Foundation for Wild Flowers & Native Plants. www.theodorepayne.org/mediawiki/index.php?title=Main_Page (site updated August 31, 2009; accessed October 27, 2011).
Cartron, Jean-Luc E., David C. Lightfoot, Jane E. Mygatt, Sandra L. Brantley, and Timothy K. Lowrey. 2008. *A Field Guide to the Plants and Animals of the Middle Rio Grande Bosque*. Albuquerque: University of New Mexico Press.
Caveney, S., D. A. Charlet, H. Freitag, M. Maier-Stolte, and A. N. Starratt. 2001. "New Observations on the Secondary Chemistry of World *Ephedra* (Ephedraceae)." *American Journal of Botany* 88:1199–208.
Center for New Crops and Plant Products. Department of Horticulture and Landscape Architecture, Purdue University. www.hort.purdue.edu/newcrop/Indices/default.html (site updated October 19, 2011; accessed October 19, 2011).
Chamberlin, Ralph V. 1912. *Memoirs of the American Anthropological Association: The Ethno-botany of the Gosiute Indians of Utah*. Memoirs of the American Anthropological Association, Volume II, Part 5 www.swsbm.com/Ethnobotany/Ethnobotany_of_Gosiute.pdf (accessed November 3, 2011).
Charters, Michael L. "California Plant Names: Latin and Greek Meanings and Derivations." www.calflora.net/botanicalnames/index.html (accessed November 3, 2011).
Cheatham, Scooter, Marshall C. Johnston, and Lynn Marshall. 1995. *Useful Wild Plants of Texas, the Southeastern and Southwestern United States, the Southern Plains, and Northern Mexico.* Vol. 1, *Abronia* through *Arundo*. Austin, TX: Useful Wild Plants.
———. 2000. *Useful Wild Plants of Texas, the Southeastern and Southwestern United States, the Southern Plains, and Northern Mexico.* Vol. 2, *Asclepias* through *Canavalia*. Austin, TX: Useful Wild Plants.
———. 2009. *Useful Wild Plants of Texas, the Southeastern and Southwestern United States, the Southern Plains, and Northern Mexico.* Vol. 3, *Canna* through *Celtis*. Austin, TX: Useful Wild Plants.
Chem-Online.org. www.chem-online.org/index.htm (accessed December 14, 2011).
Chen, Grace. 2010. "Develop a New *Lesquerella fendleri* Crop for Castor Oil Production." USDA Agricultural Research Service. www.ars.usda.gov/research/publications/publications.htm?seq_no_115=258021 (site updated November 21, 2011; accessed November 22, 2011).
Chester, Tom. 2008. "Plant Species of the South Kaibab Trail: Adonis Blazing Star, *Mentzelia multiflora*." http://tchester.org/gc/plants/species/

mentzelia_multiflora.html (site updated September 7, 2008; accessed October 19, 2011).
Colorado Native Plant Society. www.conps.org/ (accessed October 19, 2011).
Colorado Plant Database. Colorado State University Extension, Jefferson County. www.co.jefferson.co.us/coopext/ (site updated June 2010; accessed November 3, 2011).
Common and Traditionally Used Plants. San Juan River Anthropology Department, Northern Arizona University. http://home.nau.edu/sbs/anthro/sanjuanriver/plants.asp (accessed November 3, 2007).
Common Conifers in New Mexico Landscapes. New Mexico State University. http://aces.nmsu.edu/ces/forestry/documents/common-conifers-in-new-mexico-landscapes.pdf (accessed November 22, 2011).
Cornblatt, B. S., et al. 2007. "Preclinical and Clinical Evaluation of Sulforaphane for Chemoprevention in the Breast." *Carcinogenesis* 28 (7): 1485. http://carcin.oxfordjournals.org/content/28/7/1485.short (accessed November 22, 2011).
Correll, Donovan Stewart, and Marshall Conring Johnston. 1970. *Manual of the Vascular Plants of Texas*. Renner: Texas Research Foundation.
Craighead, John J., Frank C. Craighead, Jr., and Ray J. Davis. 1991. *A Field Guide to Rocky Mountain Wildflowers, Northern Arizona and New Mexico to British Columbia*. New York: Houghton Mifflin.
Cranshaw, Whitney, et al. 1994. *Insects That Feed on Colorado Trees and Shrubs*. Bulletin 506A. Colorado State University Cooperative Extension.
Davidson, Jay, and Mike Wargo. *Recognition and Control of African Rue in Nevada*. Fact Sheet FS-01-45. Cooperative Extension, University of Nevada. www.unce.unr.edu/publications/files/nr/2001/FS0145.pdf (accessed December 30, 2011).
Delange, George, and Eve Delange. "Arizona Wild Flowers." www.delange.org/ArizWFlowers/Wf.htm (accessed November 3, 2011).
———. "Xeriscape Landscape Plants and Flowers for the Arizona Desert." www.delange.org/Xeriscape/Xeriscape.htm (accessed October 27, 2011).
Desert Connections. Tucson Botanical Gardens. www.tucsonbotanical.org/desertconnections/index.php (accessed October 27, 2011).
Desert Tropicals. www.desert-tropicals.com/ (site updated April 29, 2007; accessed October 19, 2011).
Dictionary of Botanical Epithets. www.winternet.com/~chuckg/dictionary.html (site updated December 30, 2005; accessed November 3, 2011).
Dimmit, Mark A. 2000. *A Natural History of the Sonoran Desert*. Tucson,

AZ: ASDM Press. www.desertmuseumdigitallibrary.org/public/detail.php?id=ASDM19840&sp=Ferocactus%20wislizeni (accessed November 22, 2011).

Dodson, Carolyn, and William W. Dunmire. 2007. *Mountain Wildflowers of the Southern Rockies, Revealing Their Natural History*. Albuquerque: University of New Mexico.

Duke, James A. 1997. *The Green Pharmacy*. New York: St. Martin's Press.

Dunmire, William W., and Gail D. Tierney. 1995. *Wild Plants of the Pueblo Province: Exploring Ancient and Enduring Uses*. Santa Fe: Museum of New Mexico Press.

———. 1997. *Wild Plants and Native Peoples of the Four Corners*. Santa Fe: Museum of New Mexico Press.

Earle, Christopher J., ed. "*Ephedra torreyana*." The Gymnosperm Database. www.conifers.org/ep/ep/torreyana.htm (site updated November 19, 2006; accessed November 3, 2011).

Echinocereus reichenbachii. Cactus Art Nursery. www.cactus-art.biz/schede/ECHINOCEREUS/Echinocereus_reichenbachii/Echinocereus_reichenbachii/echinocereus_reichenbachii.htm (accessed November 3, 2011).

eFloras. www.efloras.org/index.aspx (accessed November 4, 2011).

Ellis, Danielle R., et al. 2004. "Production of Se-methylselenocysteine in Transgenic Plants Expressing Selenocysteine Methyltransferase." *BMC Plant Biology* 4:1. www.ncbi.nlm.nih.gov/pmc/articles/PMC343276/ (accessed October 19, 2011).

Ellis, Marion D. 2000. "Bee Tidings." University of Nebraska Cooperative Extension and the Nebraska Beekeepers Association. http://entomology.unl.edu/beekpg/tidings/btid2000/btdmay00.htm (accessed December 14, 2011).

Elpel, Thomas J. Wildflowers-and-Weeds.com. www.wildflowers-and-weeds.com/ (accessed November 3, 2011).

Emily deCamp Herbarium. Island Beach State Park, NJ. http://emilydecampherbarium.rutgers.edu (accessed November 3, 2011).

Ephedra and Ephedrine Alkaloids for Weight Loss and Athletic Performance. Fact Sheet. Office of Dietary Supplements, National Institutes of Health. http://ods.od.nih.gov/factsheets/ephedraandephedrine/ (site updated July 1, 2004; accessed December 14, 2011).

Equines and Toxic Plants. www.cnr.uidaho.edu/range/toxicplants_horses/getting_started.html (accessed December 14, 2011).

Ethnobotany of the Lower Pecos Canyonlands. Texas Beyond History. www.texasbeyondhistory.net/ethnobot/index.html (site updated May 2008; accessed November 3, 2011).

Ethnomedicinals. Plant Databases. www.ethnomedicinals.com/plntdata .html (accessed November 4, 2011).

Everitt, James H., and D. Lynn Drawe. 1993. *Trees, Shrubs & Cacti of South Texas*. Lubbock: Texas Tech University Press.

FDA Poisonous Plant Database. U.S. Food and Drug Administration. www.accessdata.fda.gov/scripts/plantox/index.cfm (site updated January 1, 2008; accessed November 5, 2011).

Fertig, Walter. 2010. "Native Plant Profiles: Four-Wing Saltbush." *Sego Lily: Newsletter of the Utah Native Plant Society* 33 (5): 10–12.

Fetrow, Charles W., and Juan R. Avila. 1999. *Professional's Handbook of Complementary and Alternative Medicines.* Springhouse, PA: Springhouse.

Flora of North America. http://fna.huh.harvard.edu/ (accessed October 18, 2011).

Food and Drugs, Substances Generally Recognized as Safe. Electronic Code of Federal Regulations, e-CFR. GPO Access. http://ecfr .gpoaccess.gov/cgi/t/text/text-idx?c=ecfr&rgn=div8&view=text &node=21:3.0.1.1.13.1.1.2&idno=21 (site updated October 31, 2011; accessed November 2, 2011).

Foster, Michael. 1892. *Bulbous Irises*. London: Royal Horticultural Society. www.ebooksread.com/authors-eng/m-michael-foster/bulbous-irises -hci/1-bulbous-irises-hci.shtml (accessed December 26, 2011).

Four Directions Institute of Native American Studies. Division of Many Circles Foundation. www.fourdir.com/?pueblo_culture.htm (accessed November 3, 2011).

Grieve, Mrs. M. 1996. *A Modern Herbal*. New York: Barnes and Noble Books.

Guarrera, Paolo M., and Leporatti M. Lucia. 2007. "Ethnobotanical Remarks on Central and Southern Italy." *Journal of Ethnobiology and Ethnomedicine* 3:23. www.ethnobiomed.com/content/3/1/23 (accessed December 26, 2011).

Gurib-Fakim, A. 2006. "*Plantago lanceolata* L." Prota: Medicinal Plants/ Plantes Médicinales. http://database.prota.org/PROTAhtml/Plantago lanceolata_En.htm (accessed October 19, 2011).

Hanson, James R. 2003. *Natural Products: The Secondary Metabolites.* Cambridge, UK: Royal Society of Chemistry.

Harmer, R. A. 1979. "Occurrence, Chemistry and Application of Betanin." *Food Chemistry* 5 (1): 81–90. www.sciencedirect.com/science/article/ pii/0308814680900667 (accessed November 22, 2011).

Hart, Charles R., Bruce B. Carpenter, and Allan McGinty, eds. 2001. *Toxic Plant Handbook: Integrated Management Strategies for West Texas.* College Station: Texas Cooperative Extension.

Henriette's Herbal. www.henriettesherbal.com/plants/index.html (accessed November 3, 2011).

Internet Sacred Text Archive. Native American Religions. www.sacred-texts.com/nam/index.htm (accessed November 4, 2011).

Jepson Flora Project: Jepson Interchange. 1993. Regents of the University of California. http://ucjeps.berkeley.edu/interchange/I_treat_indexes.html (accessed October 27, 2011).

Johns, Carl O., Lewis H. Chernoff, and Arno Viehoever. 1922. "A Saponin from Agave Lechuguilla Torrey." *Journal of Biological Chemistry* 52: 335–47. www.jbc.org/content/52/1/335.full.pdf (accessed August 19, 2012).

Kansas Wildflowers & Grasses. www.kswildflower.org/index.php (site updated July 27, 2011; accessed October 27, 2011).

Kedei, Noemi, et al. 2004. "Characterization of the Interaction of Ingenol 3-Angelate with Protein Kinase C." *Cancer Research* 64:3243–55. http://cancerres.aacrjournals.org/content/64/9/3243.full (accessed November 3, 2011).

Keeler, R. F., and D. C. Baker. 1990. "Myopathy in Cattle Induced by Alkaloid Extracts from *Thermopsis montana, Laburnum anagyroides* and a *Lupinus* sp." *Journal of Comparative Pathology* 103:169–82. www.accessdata.fda.gov/scripts/plantox/detail.cfm?id=14425 (site updated January 1, 2008; accessed December 14, 2011).

Keeler, R. F., A. E. Johnson, and R. L. Chase. 1986. "Toxicity of *Thermopsis montana* in Cattle." U.S. National Library of Medicine, National Institutes of Health. www.ncbi.nlm.nih.gov/pubmed/3698596 (accessed October 27, 2011).

Kindscher, Kelly. 1987. *Edible Wild Plants of the Prairie.* Lawrence: University Press of Kansas.

———. 1992. *Medicinal Wild Plants of the Prairie.* Lawrence: University Press of Kansas.

Kirkpatrick, Zoe Merriman. 1992. *Wildflowers of the Western Plains.* Austin: University of Texas Press.

Kish, Stacy. 2008. "*Lesquerella*: The Next Source of Biofuel." National Institute of Food and Agriculture. www.csrees.usda.gov/newsroom/impact/2008/nri/07311_lesquerella.html (site updated March 30, 2009; accessed November 22, 2011).

Kleinman, Dr. Russ. "Vascular Plants of the Gila Wilderness." Department of Natural Sciences and the Dale A. Zimmerman Herbarium, Western New Mexico University. www.wnmu.edu/academic/nspages/gilaflora/index.html (accessed October 18, 2011).

Kupchan, S. M., J. M. Cassady, J. Bailey, and J. R. Knox. 1965. "Tumor Inhibitors XII: Gaillardin, a New Cytotoxic Sesquiterpene Lactone from *Gaillardia pulchella*." *Journal of Pharmaceutical Sciences* 54:1703–05. doi:10.1002/jps.2600541142.

Kwiecinski, Tristan. 2005. *Salmon Ruins Trail Guide*. Bloomfield, NM: San Juan County Museum Association.

Ladyman, Juanita A. "*Opuntia leptocaulis* DC." www.fs.fed.us/global/iitf/pdf/shrubs/Opuntia leptocaulis.pdf (accessed November 2, 2011).

Lira, Rafael, et al. 2009. "Traditional Knowledge and Useful Plant Richness in the Tehuacan-Cuicatlan Valley, Mexico." *Economic Botany* 63 (3): 271–87.

Lowengard, Sarah. 2006. *The Creation of Color in Eighteenth-Century Europe*. New York: Columbia University Press e-book. www.gutenberg-e.org/lowengard/index.html (accessed December 30, 2011).

Lucero, Mary, et al. 2006. "Volatile Composition of *Guiterrizia sarothrae* (Broom Snakeweed) as Determined by Steam Distillation and Solid Phase Microextraction." *Journal of Essential Oil Research* 18:121–25.

Maheshwari, J. K. "Processing and Utilization of Perennial Vegetation in the Arid-Zone of India." www.fao.org/docrep/T0115E/T0115E0g.htm (accessed October 18, 2011).

Martin, William C., and Charles R. Hutchins. 1984. *Spring Wildflowers of New Mexico.* Albuquerque: University of New Mexico Press.

Marx, Ernie. "Eastern Colorado Wildflowers." www.easterncoloradowildflowers.com/ (accessed November 2, 2011).

McArthur, E. Durant. 1977. "Environmentally Induced Changes of Sex Expression in Atriplex canescens." *Heredity* 38 (1): 97–103.

McLain-Romero, J., R. Creamer, H. Zepeda, J. Strickland, and G. Bell. 2004. "The Toxicosis of *Embellisia* Fungi from Locoweed (*Oxytropis lambertii*) Is Similar to Locoweed Toxicosis in Rats." *Journal of Animal Science* 82:2169–74. www.animal-science.org/content/82/7/2169.full (accessed December 14, 2011).

Medical Alternative Medicine Index. University of Maryland Medical Center. www.umm.edu/altmed/ (accessed November 4, 2011).

Medline Plus. U.S. National Library of Medicine. www.nlm.nih.gov/medlineplus/ (site updated December 14, 2011; accessed December 14, 2011).

Mellado, M., and A. Olvera. 2008. "Diets of Prairie Dogs (*Cynomys mexicanus*) Co-existing with Cattle or Goats." *Mammalian Biology* 73: 33–39. www.gprc.org/Mellado.pdf (accessed November 2, 2011).

Miller, George Oxford. 2007. *Landscaping with Native Plants of the Southwest*. Minneapolis: Voyageur Press.

Misra, P., et al. 2008. "Antileishmanial Activity Mediated by Apoptosis and Structure-Based Target Study of Peganine Hydrochloride Dihydrate: An Approach for Rational Drug Design." *Journal of*

Antimicrobial Chemotherapy 52 (5): 998–1002. www.ncbi.nlm.nih.gov/pubmed/18694906 (accessed December 30, 2011).

Missouri Plants. www.missouriplants.com/index.html (site updated February 8, 2007; accessed November 2, 2011).

Moerman, Daniel E. 2009. *Native American Medicinal Plants: An Ethnobotanical Dictionary*. Portland, OR: Timber Press.

Mondragon-Jacobo, C. 1999. "Low Input Agricultural Systems Based on Cactus Pear for Subtropical Semiarid Environments." In *Perspectives on New Crops and New Uses*, 367–72. Alexandria, VA: ASHS Press. www.hort.purdue.edu/newcrop/proceedings1999/v4-367.html (accessed November 22, 2011).

Mondragon-Jacobo, C., and S. Perez-Gonzalez. 1996. "Native Cultivars of Cactus Pear in Mexico." In *Progress in New Crops*, 446–50. Arlington, VA: ASHS Press. www.hort.purdue.edu/newcrop/proceedings1996/V3-446.html (accessed November 22, 2011).

Moore, Michael. 1979. *Medicinal Plants of the Mountain West*. Santa Fe: Museum of New Mexico Press.

———. 1989. *Medicinal Plants of the Desert and Canyon West*. Santa Fe: Museum of New Mexico Press.

———. 1990. *Los Remedios: Traditional Herbal Remedies of the Southwest.* Santa Fe: Red Crane Books.

Müller, C., N. Agerbirk, C. E. Olsen, J. L. Boevé, U. Schaffner, and P. M. Brakefield. 2001. "Sequestration of Host Plant Glucosinolates in the Defensive Hemolymph of the Sawfly *Athalia rosae*." *Journal of Chemical Ecology* 27 (12): 2505–16. doi:10.1023/A:1013631616141. PMID 11789955. http://link.springer.com/article/10.1023%2FA%3A1013631616141¿LI=true#page-1 (accessed November 22, 2011).

Native American Ethnobotany. A Database of Foods, Drugs, Dyes and Fibers of Native American Peoples, Derived from Plants. University of Michigan, Dearborn. http://herb.umd.umich.edu/ (accessed November 3, 2011).

Native Seed Network. www.nativeseednetwork.org/index (accessed December 26, 2011).

Natural Medicinal Herbs. www.naturalmedicinalherbs.net/ (accessed October 27, 2011).

Navajo Ethnobotany. Dykeman Roebuck Archaeology. www.drarchaeology.com/herbal/navajoethnobot.htm (accessed November 4, 2011).

Nazaruk, J., et al. 2010. "In Vitro Antiproliferative and Antifungal Activity of Essential Oils from *Erigeron acris* L. and *Erigeron annuus* (L.) Pers." *A Naturforsch* 65 (11–12): 642–46.

NCI Drug Dictionary. National Cancer Institute at the National Institutes of Health. www.cancer.gov/ (accessed November 3, 2011).

Niethammer, Carolyn. 1974. *American Indian Food and Lore: 150 Authentic Recipes*. New York: Collier.

———. 1987. *The Tumbleweed Gourmet: Cooking with Wild Southwestern Plants*. Tucson: University of Arizona Press.

———. 2004. *Prickly Pear Cookbook*. Tucson, AZ: Rio Nuevo.

New Mexico Flores. http://newmexicoflores.com/index.html (accessed October 18, 2011).

New Mexico Rare Plants. New Mexico Rare Plant Technical Council. http://nmrareplants.unm.edu/rarelist_single.php?SpeciesID=85 (accessed October 19, 2011).

Non-Wood Forest Products from Conifers. 1998. Forestry Department, Food and Agriculture Organization of the United Nations. www.fao.org/docrep/x0453e/X0453e00.htm (accessed December 26, 2011).

Norman, James O., et al. 1976. "Effects of Sesquiterpene Lactones on the Growth of *Bacillus Thuringiensis*." *Antimicrobial Agents and Chemotherapy* 9 (3): 535–39.

Northern Arizona Flora. A Photographic, Annotated Catalog of Northern Arizona Vascular Plants. www.nazflora.org/index.html (site updated May 28, 2011; accessed November 3, 2011).

Northern Bushcraft. http://northernbushcraft.com/ (accessed October 19, 2011).

O'Connell, Mary. "Medicinal Plants of the Southwest." Department of Agronomy & Horticulture, New Mexico State University. http://medplant.nmsu.edu (site updated February 13, 2008; accessed October 18, 2011).

Ogbourne, Steven Martin, et al. 2009. "Method for Wound Healing." Patent Application. www.faqs.org/patents/app/20090215884 (accessed December 31, 2011).

———. 2010. "Treatment of Virally Induced Lesions." Patent Application. www.freepatentsonline.com (accessed December 31, 2011).

Oliggo-Fiber Inulin. Cargill. www.cargill.com/food/na/en/products/health-promoting-ingredients/oliggo-fiber-inulin/index.jsp (accessed November 3, 2011).

Opuntia lindheimeri var. *linguiformis*, Cow's Tongue Prickly Pear. Master Gardeners of the University of Arizona Pima County Cooperative Extension. http://ag.arizona.edu/pima/gardening/aridplants/Opuntia_lindheimeri.html (site updated March 16, 2006; accessed November 3, 2011).

Opuntia spinosior. The Plant List. AZArboretum.org. www.azarboretum.org/plantlist/canecholla.html (accessed November 2, 2011).

Orwa C., A. Mutua, R. Kindt, R. Jamnadass, and A. Simons. 2009. "*Melia azedarach*, Persian Lilac." Agroforestry Database: A Tree Reference and Selection Guide. www.worldagroforestry.org/af/treedb/ (accessed October 19, 2011).

Panter, Kip E., D. R. Gardner, S. T. Lee, J. A. Pfister, M. H. Ralphs, B. L. Stegelmeier, and L. F. James. "Important Poisonous Plants of the United States." In *Veterinary Toxicology*, 825–72. http://ddr.nal.usda.gov/bitstream/10113/9232/1/IND43946780.pdf (accessed December 14, 2011).

Pfeiffer, F. A., and M. C. Calhoun. 1987. "Effects of Environmental, Site and Phenological Factors on Hymenoxon Content of Bitterweed (*Hymenoxys odorata*)." *Journal of Animal Science* 65: 1553–62.

Phoenix, G. K., and M. C. Press. 2005. "Linking Physiological Traits to Impacts on Community Structure and Function: The Role of Root Hemiparasitic Orobanchaceae." *Journal of Ecology* 93 (1): 67–78.

The Plant List. 2010. http://222.theplantlist.org/ (accessed October 19, 2011).

Plant Profiler. SIGMA Life Science, Sigma-Aldrich. www.sigmaaldrich.com/life-science/nutrition-research/learning-center/plant-profiler.html (accessed November 5, 2011).

Plants Database. USDA Natural Resources Conservation Service. http://plants.usda.gov/java/ (site updated October 24, 2011; accessed November 2, 2011).

Plants for a Future: Species Database. 1996–2010. www.pfaf.org (accessed October 18, 2011).

Plants of California: A Guide to Useful, Edible and Medicinal Plants of California. www.plantsofcalifornia.com/ (accessed October 19, 2011).

Plants of Wisconsin. Robert W. Freckmann Herbarium, University of Wisconsin, Stevens Point. http://wisplants.uwsp.edu/WisPlants.html (accessed November 2, 2011).

Poisoning by Locoweeds. Mike Sanderson's Lab, Department of Ecology and Evolutionary Biology, University of Arizona. http://loco.biosci.arizona.edu/astragalus/locoweedspp.html (accessed November 2, 2011).

Poisonous Plants of Pennsylvania. Department of Agriculture, Commonwealth of Pennsylvania. http://cal.vet.upenn.edu/projects/poison/agbook/aghome.htm (accessed December 26, 2011).

Potentilla (Wild Agrimony). Cosmetochem. www.specialchem4cosmetics.com/common/cos/services/ingredients/displayproduct.aspx?id=11449 (accessed November 3, 2011).

Poulton, Jonathan E. 1990. "Cyanogenesis in Plants." *Plant Physiology*

94: 401–5. www.ncbi.nlm.nih.gov/pmc/articles/PMC1077245/pdf/plntphys00811-0019.pdf (accessed October 18, 2011).

Prairie Smoke Geum. www.paghat.com/geum.html (accessed November 3, 2011).

Pratt, Mindy, Jim Bowns, Roger Banner, and Allen Rasmussen. "Range Plants of Utah." Utah State University. http://extension.usu.edu/range/FORB.HTM (site updated March 2004; accessed November 3, 2011).

PubMed Health. U.S. National Library of Medicine. www.ncbi.nlm.nih.gov/pubmedhealth/PMH0000099/ (site updated November 15, 2011; accessed December 14, 2011).

Quinn, Colin F., et al. 2007. "The Role of Selenium in Protecting Plants against Prairie Dog Herbivory: Implications for the Evolution of Selenium Hyperaccumulation." Great Plains Restoration Council. www.gprc.org?Quinn.pdf (accessed November 2, 2011).

———. 2010. "Selenium Hyperaccumulation Offers Protection from Cell Disruptor Herbivores." *BMC Ecology* 10:19. doi:10.1186/1472-6785-10-19. www.biomedcentral.com/1472-6785/10/19 (accessed November 2, 2011).

Rainey, Katharine D., and Karen R. Adams. 2004. "Plant Use by Native Peoples of the American Southwest: Ethnographic Documentation. Compendium A: Ethnographically Documented Uses of Plants." www.crowcanyon.org/ResearchReports/Archaeobotanical/Plant_Uses/compendium_A.asp (accessed November 3, 2011).

Range Plants of Utah. Utah State University. http://extension.usu.edu/range/index.htm (site updated March 2004; accessed October 27, 2011).

Ratzka, A. 2002. "Disarming the Mustard Oil Bomb." *Proceedings of the National Academy of Sciences* 99 (17): 11223. doi:10.1073/pnas.172112899. www.pnas.org/content/99/17/11223.long (accessed November 22, 2011).

Red Butte Canyon Research Natural Area. http://redbuttecanyon.net/flora.html (accessed October 27, 2011).

Reed, Paul F., ed. 2006. *Thirty-Five Years of Archaeological Research at Salmon Ruins, New Mexico.* Vol. 3, *Archaeobotanical Research and Other Analytical Studies.* Tucson, AZ: Center for Desert Archaeology.

REEIS (Research, Education & Economics Information System). U.S. Department of Agriculture. www.reeis.usda.gov/web/crisprojectpages/196306.html (accessed November 3, 2011).

Remington, Joseph P., and Horatio C. Wood, et al., eds. 1918. "How to Use Herbs with the Dispensatory of the United States of America." www.havaris.ca/info/017/index.html (accessed November 3, 2011).

Roalson, Eric H., and Kelly W. Allred. 1997. "Atlas of New Mexico *Astragalus* and *Oxytropis*." New Mexico State University Agricultural Experiment Station Research Report 715. http://aces.nmsu.edu/pubs/research/livestock_range/RR715.pdf (accessed November 3, 2011).

Roberson, James Ray. 1969. *Manufacturing and Marketing a Mexican-Type Cheese in West Texas*. http://etd.lib.ttu.edu/theses/available/etd-05122009-31295004136916/unrestricted/31295004136916.pdf (accessed October 18, 2011).

Rocky Mountain Arsenal National Wildlife Refuge. U.S. Fish and Wildlife Service. www.fws.gov/rockymountainarsenal/overview/overview.htm (site updated August 9, 2011; accessed October 27, 2011).

Rook, Earl J. S. 2002. "*Equisetum laevigatum*, Smooth Scouring Rush." Rook.org. www.rook.org/earl/bwca/nature/ferns/equisetumlae.html (site updated February 26, 2004; accessed November 2, 2011).

Royo, A. R. "Cholla Cactus." DesertUSA. www.desertusa.com/flora.html (accessed November 2, 2011).

Roys, Ralph L. 1939. *The Ethno-botany of the Maya*. Philadelphia: Institute for the Study of Human Issues.

Salywon, Andrew M., et al. 2005. "Evaluation of New *Lesquerella* and *Physaria* (Brassicaceae) Oilseed Germplasm." *American Journal of Botany* 92:53–62. www.amjbot.org/content/92/1/53.full (accessed November 22, 2011).

San Juan Heritage Plants. San Juan School District, NM. http://dine.sanjuan.k12.ut.us/heritage/land/plants/plants.htm (site updated April 2003; accessed November 3, 2011).

Savinelli, Alfred. 1997. *Plants of Power.* Taos, NM: Alfred Savinelli.

Scher, Julia. "Federal Noxious Weed Disseminules of the U.S." http://keys.lucidcentral.org/keys/FNW/FNW%20Disseminules%20Key/html/index.htm (accessed November 2, 2011).

Scottish Thistle. ScotClans. www.scotclans.com/scottish_myths/symbolic_scotland/thistle.html (accessed November 5, 2011).

Scully, Richard W. 2010. "Solanaceae of Colorado." www.conps.org/pdf/Workshops/Solan web version pt.1.pdf (accessed November 4, 2011).

SEINet (Southwest Environmental Information Network). http://swbiodiversity.org/seinet/imagelib/index.php (accessed November 3, 2011).

Shaw, Joe, and Dave Ferguson. "Opuntiads of the USA." http://opuntiads.com/index.html (accessed November 2, 2011).

Silverthorne, Elizabeth. 1996. *Legends and Lore of Texas Wildflowers*. College Station: Texas A&M University Press.

Simpson, Benny J. 1988. *A Field Guide to Texas Trees*. Austin: Texas Monthly Press.

Singh, K. N., and Rajesh Kaushal. 2007. "Comprehensive Notes on Commercial Utilization, Characteristics and Status of Steroid Yielding Plants in India." www.ethnoleaflets.com/leaflets/steroids.htm (accessed October 18, 2011).

Smith, James E. "Bud." 2009. "Engelmann's Daisy, *Engelmannia peristenia* (Raf.)" USDA Natural Resources Conservation Service. www.plant-materials.nrcs.usda.gov/pubs/txpmcfs9177.pdf (accessed October 27, 2011).

Society of Economic Botany. www.econbot.org/_welcome_/to_seb.php (accessed November 4, 2011).

Sonoran Desert Naturalist. www.arizonensis.org/sonoran/ (site updated August 28, 2011; accessed October 18, 2011).

Southeastern Arizona Wildflowers and the Plants of the Sonoran Desert. www.fireflyforest.com/flowers/ (accessed October 19, 2011).

Srinibas Das, Amrish Kumar Tyagi, and Harjit Kaur. 2000. "Cancer modulation by glucosinolates: A review." *Current Science* 79 (12): 1665. www.ias.ac.in/currsci/dec252000/1665.pdf (accessed November 22, 2011).

Stegelmeier, Bryan L., et al. 2010. "Experimental Rayless Goldenrod (*Isocoma pluriflora*) Toxicosis in Goats." *Journal of Veterinary Diagnostic Investigation* 22:570–77.

Stevens County Noxious Weed Control Board. www.co.stevens.wa.us/weedboard/index.htm (site updated December 1, 2011; accessed December 14, 2011).

Stintzing, F. C., A. Schieber, and R. Carle. 1999. "Amino Acid Composition and Betaxanthin Formation in Fruits from *Opuntia ficus-indica*." *Planta Medica* 65 (7): 632–35. http://cat.inist.fr/?aModele=afficheN&cpsidt=1976337 (accessed November 22, 2011).

Stuart, B. P., R. J. Cole, and H. S. Gosser. 1981. "Cocklebur (*Xanthium strumarium* L. var. *strumarium*) Intoxication in Swine: Review and Redefinition of the Toxic Principle." *Veterinary Pathology* 18 (3): 368–83.

Strunk, Maya. 2001. "Medicinal Plants of the Southwest: *Ephedra viridis*." New Mexico State University. http://medplant.nmsu.edu/ephedra.html (site updated February 13, 2008; accessed November 2, 2011).

Substances Generally Recognized as Safe. Spices and Other Natural Seasonings and Flavorings. Code of Federal Regulations. Title 21: Food and Drugs. http://cfr.vlex.com/vid/182-spices-natural-seasonings-flavorings-19706866 (accessed November 3, 2011).

Sullivan, Preston. 2004. "Field Bindweed Control Alternatives." NCAT Sustainable Agriculture Project. https://attra.ncat.org/ (accessed November 22, 2011).

Summary of Color Additives for Use in United States in Foods, Drugs, Cosmetics, and Medical Devices. U.S. Food and Drug Administration. www.fda.gov/ForIndustry/ColorAdditives/ColorAdditiveInventories/ucm115641.htm (site updated October 20, 2011; accessed November 22, 2011).

Temple College Landscape. www.templejc.edu/Campus/landscape/landscape.htm (site updated October 16, 2011; accessed October 18, 2011).

Thie, Krista K. 1999. *Medicinal Plants of the Pacific Northwest: A Digest of Anthropological Writings about Native American Uses*. White Salmon, WA: Longevity Herb Press.

Tilley, Derek J. 2011. "Propagation Protocol for Production of *Packera multilobata* (Torr. & A. Gray ex A. Gray) W. A. Weber & A. Love Seeds." USDA Natural Resources Conservation Service, Aberdeen Plant Materials Center, Aberdeen, ID. Native Plant Network. www.nativeplantnetwork.org/Network/ViewProtocols.aspx?ProtocolID=3880 (accessed November 3, 2011).

———. 2011. "Request for Seed Collections: Lobeleaf Groundsel, Packera multilobata (Torr. & A. Gray ex A. Gray) W. A. Weber & A. Love." USDA Natural Resources Conservation Service. www.plant-materials.nrcs.usda.gov/pubs/idpmcot10220.pdf (accessed November 5, 2011).

Tous, J., and L. Ferguson. 1996. "Mediterranean Fruits." In *Progress in New Crops*, 416–30. Arlington, VA: ASHA Press. www.hort.purdue.edu/newcrop/proceedings1996/V3-416.html#Cactus%20Pear (accessed November 22, 2011).

Toxic Agents in Plants. Plants Poisonous to Livestock. Department of Animal Science, Cornell University. www.ansci.cornell.edu/plants/toxicagents/index.html (site updated August 19, 2009; accessed November 3, 2011).

Toxic Plants of Texas. AgriLife Extension Service, Texas A&M System. http://essmextension.tamu.edu/plants/toxics/ (accessed October 18, 2011).

Tree New Mexico. New Mexico Tree Species Information. www.treenm.com/education/species.shtml (accessed November 4, 2011).

Turner, Billie L., H. Nichols, G. Denny, and O. Doron. 2003. *Atlas of the Vascular Plants of Texas*. Fort Worth: Botanical Research Institute of Texas.

University of Michigan Herbarium. www.lsa.umich.edu/herb/ (accessed November 3, 2011).

USDA Forest Service. Fire Effects Information System. www.fs.fed.us/database/feis/index.html (updated September 29, 2011; accessed October 18, 2011).

Van Dijk, Peter J. 2003. "Ecological and Evolutionary Opportunities of Apomixis: Insights from *Taraxacum* and *Chondrilla*." Royal Society, London. www.ncbi.nlm.nih.gov/pmc/articles/PMC1693208/pdf/12831477.pdf (accessed November 10, 2011).

Van Niekerk, W. A., et al. 2004. "Interspecies and Location Variation in Oxalic Acid Concentrations in Certain *Atriplex* Species and *Cassia sturtii*." *South African Journal of Animal Science* 34:101–4.

Virginia Tech Weed Identification Guide. www.ppws.vt.edu/scott/weed_id/ephde.htm (accessed November 2, 2011).

Vogel, Virgil. 1970. *American Indian Medicine*. Norman: University of Oklahoma Press.

Wallens, Celso Olmedo Bejarano, et al. 2009. "Quick Corn Nixtamalization Process." Patent Application. www.faqs.org/patents/app/20090291186 (accessed December 31, 2011).

Walsh, Guillermo Cabrera, and Robert F. Schroder. 1999. "TAIUIA—Corn Rootworms Just Can't Get Enough." *Agricultural Research* (April). www.ars.usda.gov/is/AR/archive/apr99/tai0499.htm?pf=1 (site updated November 21, 2011; accessed November 22, 2011).

Warnock, Barton H. 1970. *Wildflowers of the Big Bend Country, Texas*. Alpine, TX: Sul Ross State University.

Weed Science Society of America. www.wssa.net/Weeds/ID/WorldOfWeeds.htm#x (site updated January 13, 2007; accessed November 3, 2011).

Weekley, Ernest. 1921. *An Etymological Dictionary of Modern English*. London: J. Murray.

Whitson, Tom D., Larry C. Burrill, Steven A. Dewey, David W. Cudney, B. E. Nelson, Richard D. Lee, and Robert Parker, eds. 2008. *Weeds of the West*. 10th ed. Las Cruces, NM: Western Society of Weed Science.

Wildflowers, Ferns & Trees of Colorado, New Mexico, Arizona & Utah. www.swcoloradowildflowers.com/ (accessed October 19, 2011).

Wilson, Hugh D. 1981. "Domesticated *Chenopodium* of the Ozark Bluff Dwellers." *Economic Botany* 35 (2): 233–39.

———. 1990. "*Quinua* and Relatives." *Economic Botany* 44 (3 Suppl.): 92–110.

Wilson, Paul, Andrea D. Wolfe, W. Scott Armbruster, and James D. Thomson. 2007. "Constrained Lability in Floral Evolution: Counting Convergent Origins of Hummingbird Pollination in *Penstemon* and *Keckiella*." *New Phytologist* 176:883–90. http://onlinelibrary.wiley.com/doi/10.1111/j.1469-8137.2007.02219.x/full (accessed December 26, 2011).

Winter, Ruth. 1978. *A Consumer's Dictionary of Food Additives*. New York: Crown.

Wittstock, U., N. Agerbirk, E. J. Stauber, C. E. Olsen, M. Hippler, T. Mitchell-Olds, J. Gershenzon, and H. Vogel. 2004. "Successful Herbivore Attack Due to Metabolic Diversion of a Plant Chemical Defense." *Proceedings of the National Academy of Sciences* 101 (14): 4859–64. doi:10.1073/pnas.0308007101. PMC 387339. PMID 15051878. www.pubmedcentral.nih.gov (accessed November 22, 2011).

Wolfe, Andrea D. "The Penstemon Website: Home of the Beardtongue." www.biosci.ohiostate.edu/~awolfe/Penstemon/spp/Pcrandallii.html (site updated August 16, 2001; accessed October 18, 2011).

Wolfe, Andrea D., Christopher P. Randle, Shannon L. Datwyler, Jeffery J. Morawetz, Nidia Arguedas, and Jose Diaz. 2006. "Phylogeny, Taxonomic Affinities and Biogeography of *Penstemon* (Plantaginaceae) Based on ITS and cpDNA Sequence Data." *American Journal of Botany* 93:1699–713. www.amjbot.org/content/93/11/1699.full (accessed December 26, 2011).

The World's Healthiest Foods. The George Mateljan Foundation. www.whfoods.org (accessed November 10, 2011).

Worthington, Dr. Richard D. 2003. "Preliminary Inventory of the Flora of White Sands National Monument, New Mexico." New Mexico State University. http://cahe.nmsu.edu/academics/rangescienceherbarium/domcuments/wsnm.pdf (accessed November 4, 2011).

Wyman, Leland C., and Stuart K. Harris. 1941. "Navajo Indian Medical Ethnobotany." *University of New Mexico Bulletin*. Anthropological Series, vol. 3, no. 5. Albuquerque: University of New Mexico Press.

Yetman, David. 2009. *50 Common Edible & Useful Plants of the Southwest*. Tucson, AZ: Western National Parks Association.

Yu, Larina S. "Weeds—*Cardaria draba* (L.) Desv.—Heart-Podded Hoary Cress." Interactive Agricultural Ecological Atlas of Russia and Neighboring Countries: Economic Plants and Their Diseases, Pests and Weeds. www.agroatlas.ru/en/content/weeds/Cardaria_draba/ (accessed November 3, 2011).

Page numbers in **bold** refer to each plant's main entry.

ABOUT THE AUTHORS

Willa F. Finley grew up riding horses and herding cows in the native pastures around her home in Floydada, Texas, where buffalo grass, broomweed, and milkweed were part of the natural flora. Her affinity for all things botanical was passed down through her maternal grandmother and her mother, both of whom had extensive flower gardens. Since earning advanced degrees in plant systematics and plant genetics, she has worked in agriculture research and in agribusiness in the Middle East, North Africa, and Central Africa, and for many years was senior researcher in agricultural economics for LMC International in Oxford, UK. She is now involved in agricultural development projects in Africa, helping poor African women farmers improve their economic status. Her photography has received international recognition. She is available for talks about wildflowers as well as about her work in Africa.
www.landofenchantmentwildflowers.com

LaShara Nieland, a native of Big Spring, Texas, has always been fascinated by the natural world, as well as by history. Her father, Charlie Shanks, passed along his interest in and respect for Native Americans to LaShara during many trips to New Mexico throughout her childhood. During her years at Abilene Christian University, where she earned two degrees, she began studying how ancient people used wild plants for food and medicine, an interest that continues today. She incorporated this research into her lessons as she taught biology in Abilene and in Odessa, Texas, for twenty-seven years. Now retired, she still lives in Odessa with her husband, Andy, and continues to photograph wildflowers, research their uses, and give talks to various groups.
www.landofenchantmentwildflowers.com